NUREG–1335

I0488125

Individual Plant Examination: Submittal Guidance

Final Report

Manuscript Completed: July 1989
Date Published: August 1989

Office of Nuclear Regulatory Research
Office of Nuclear Reactor Regulation
U.S. Nuclear Regulatory Commission
Washington, DC 20555

ABSTRACT

Based on a Policy Statement on Severe Accidents Regarding Future Designs and Existing Plants, the performance of a plant examination is requested from the licensee of each nuclear power plant. The plant examination looks for vulnerabilities to severe accidents and cost-effective safety improvements that reduce or eliminate the important vulnerabilities. This document delineates the guidance for reporting the results of that plant examination.

TABLE OF CONTENTS

LIST OF TABLES

ACKNOWLEDGMENTS

This document represents the staff position on the Individual Plant Examination process. Representatives of both the Office of Nuclear Regulatory Research and the Office of Nuclear Reactor Regulation were active contributors to the process; they are named below. In addition, significant input was received from contractors to the NRC, who are also named below, especially in the preparation of early drafts. Louise Gallagher, of the NRC, provided technical editing.

NRC

Richard Barrett
William Beckner
Franklin Coffman
Frank Congel
Thomas Cox
Adel El-Bassioni
Farouk Eltawila
John Flack
R. Wayne Houston
Glenn Kelly
Jocelyn Mitchell
Robert Palla
Mark Rubin
Themis Speis
Charles Tinkler
Ashok Thadani

Contractors

James Meyer (SCIENTECH, Inc.)
Mohamad Modarres (University of Maryland)
Trevor Pratt (Brookhaven National Laboratory)
Theofanis Theofanous (University of California)

1. INTRODUCTION AND OBJECTIVES

1.1 Background

On August 8, 1985, NRC issued a Policy Statement on Severe Accidents Regarding Future Designs and Existing Plants (50 FR 32138) that introduced the Commission's plan to address severe accident issues for existing commercial nuclear power plants. (The staff in a separate effort is developing recommendations on the treatment of severe accident issues for future LWRs.) Over the past several years, the Commission has developed an approach to implement this plan for existing plants and recently has issued a Generic Letter (Ref. 1) that communicates this plan to all utilities. Each licensed nuclear power plant is requested to perform a plant examination that looks for vulnerabilities to severe accidents and cost-effective safety improvements that reduce or eliminate the important vulnerabilities. The specific objectives for these Individual Plant Examinations (IPEs) are for each utility to (1) develop an overall appreciation of severe accident behavior; (2) understand the most likely severe accident sequences that could occur at its plant; (3) gain a more quantitative understanding of the overall probability of core damage and radioactive material releases; and (4) if necessary, reduce the overall probability of core damage and radioactive material release by appropriate modifications to procedures and hardware that would help prevent or mitigate severe accidents. Upon completion of the examination, the utility will be required to submit a report to NRC describing the results and conclusions of the examination. This submittal will be reviewed and evaluated by the NRC.

This IPE submittal guidance document establishes format and content guidelines for the utility submittals. There are NRC and industry reports that help to put this document into proper perspective and help to give background for many of the specific matters presented herein.

- "Severe Accident Insights Report," NUREG/CR-5132 (Ref. 2). This report describes the conditions and events that nuclear power plant personnel may encounter during the latter stages of a severe core damage accident and what the consequences might be of actions they may take during these latter stages. The report also describes what can be expected of the performance of the key barriers to fission product release (primarily containment systems), what decisions the operating staff may face during the course of a severe accident, and what could result from these decisions based on our current state of knowledge of severe accident phenomena.

- "Assessment of Severe Accident Prevention and Mitigation Features," NUREG/CR-4920, Volumes 1-5 (Ref. 3). This series of reports describes plant features and operator actions found to be important in either preventing or mitigating severe accidents in LWRs with five different types of containments.

- "PRA Procedures Guide," NUREG/CR-2300 (Ref. 4). This report is a guide to the performance of probabilistic risk assessments (PRAs) for nuclear power plants.

- "PRA Review Manual," NUREG/CR-3485 (Ref. 5). This report describes an approach for reviewing a Level 1 type PRA (a PRA that carries the accident analysis up to the point of calculating the probability of core damage or core melt).

- "Probabilistic Safety Analysis Procedures Guide," NUREG/CR-2815 (Ref. 6). This report provides the structure of a probabilistic safety study that is to be performed and indicates which products of the study are valuable for regulatory decisionmaking.

- "Individual Plant Evaluation Methodology for LWRs," IDCOR (Ref. 7). This industry report provides, in a BWR volume and a PWR volume, methodology for plant-specific evaluation of the probability of severe accidents.

- "Staff Evaluation of the IDCOR IPEM for PWRs," "Staff Evaluation of the IDCOR IPEM for BWRs" (Ref. 8). These two reports describe the enhancements to the front-end of the Individual Plant Examination Methodology (IPEM) that the staff considers necessary before the front-end IPEM should be used for an IPE.

- "Evaluation of System Interactions in Nuclear Power Plants," NUREG-1174 (Ref. 9). This report presents a summary of the activities related to Unresolved Safety Issue (USI) A-17, "System Interactions in Nuclear Power Plants," and includes the NRC staff's conclusions based on those activities. Of particular importance is the discussion of internal flooding, including water intrusion.

- "Accident Sequence Evaluation Program--Human Reliability Analysis Procedure," NUREG/CR-4772 (Ref. 10). This document describes the human reliability analysis method used in the NUREG-1150 assessment.

- "Recovery Actions in PRA for the Risk Method Integration and Evaluation Program," NUREG/CR-4834, Volumes 1 and 2 (Ref. 11). These two volumes describe an improved method for estimating recovery actions that are based upon observable data rather than expert judgment. The method was applied to selected recovery actions and provides guidance on the application of the method to other human actions.

- "Comparison and Application of Quantitative Human Reliability Analysis Methods for Risk Method Integration and Evaluation Program," NUREG/CR-4835 (Ref. 12). This document is a topical survey of available methods that amplifies on the material provided in Reference 6.

1.2 Purpose

The purpose of this document is to provide format and content guidelines for the utility submittals. The reasons for having these guidelines are to provide sufficient submittal content for an effective review and to provide a format that allows for an efficient and consistent submittal review. This document should be used by the utilities as they perform their IPEs and prepare their submittal reports.

In addition, the appendices to this document contain: (1) an approach to the back-end portion of a PRA, (2) references to PRAs performed or reviewed by the NRC, (3) NRC responses to questions and comments raised at the IPE workshop, and (4) staff review guidance. This additional information should be useful in performing the IPEs.

1.3 Scope

The scope of this report is consistent with the IPE program as outlined in the Generic Letter (Ref. 1). This report presents submittal guidance for the IPEM and the PRA method of performing an IPE. These are the first two of the three options discussed in the Generic Letter. (The third option, that of choosing some other method (unspecified), will be treated on a case-by-case basis as necessary.) It should also be noted that the IPE program stops with the radionuclide release characterization. The IPE should carry through evaluation of the behavior of the containment and radionuclide releases to enable utility personnel to understand these phenomena and to provide a basis for the development of an accident management capability. Finally, this document makes no substantive distinction between the two IPE options, namely, the IPEM by IDCOR (Ref. 7) and PRAs, in the submittal guidelines. All limitations of the IPEM and enhancements to the front-end IPEM for use in the IPE program are delineated in the staff evaluation reports (Ref. 8). Therefore, they are not repeated in this document.

1.4 Goals

This document is to provide a uniform mechanism for allowing the NRC staff to draw conclusions regarding the implementation of the Severe Accident Policy Statement for existing plants.

- The NRC staff will want to determine whether the IPE has achieved the objectives of the IPE program. Specifically, as stated in the Generic Letter, "The NRC will evaluate licensee IPE submittals to obtain reasonable assurance that the licensee has adequately analyzed the plant design and operations to discover instances of particular vulnerability to core melt or unusually poor containment performance given a core melt accident. Further, the NRC will assess whether the conclusions the licensee draws from the IPE regarding changes to the plant systems, components, or accident management procedures are adequate. The consideration will include both quantitative measures and nonquantitative judgment." A positive staff conclusion would be that there is a likelihood that the IPEM or the PRA represents the plant and its operation and that it had the capability to identify previously unrecognized vulnerabilities. It could then be concluded that the utility was, or will be, on firm ground when making improvements and implementing an effective accident management program.

- The basis for the request in the Generic Letter (Ref. 1) for involvement of utility staff in the IPE review is the belief that the maximum benefit from the performance of an IPE would be realized if the utility's staff were involved in all aspects of the examination and that involvement would facilitate integration of the knowledge gained from the examination into emergency operating procedures and training programs.

1.5 Review Process for This Document

This document was issued in draft form in January 1989. A workshop with interested members of the public was held in Fort Worth, Texas, on February 28 and March 1 and 2, 1989. Written comments were received from 11 parties. These comments and comments from the workshop (based on a review of the transcript) have been considered in developing this document in final form. Appendix C contains the comments and the NRC staff responses to them.

2. SUBMITTAL GUIDELINES: FORMAT AND CONTENT

This section provides the format and content guidelines for the utility submittals. The major parts of this section are the front-end analysis (Section 2.1), the back-end analysis (Section 2.2), unique safety features and plant improvements (Section 2.3), and the utility team (Section 2.4). The utilities are requested to submit their IPE reports using the standard table of contents given in Table 2.1. This will facilitate review by the NRC and promote consistency among various submittals. The content of the elements of this Table of Contents is discussed in Sections 2.1, 2.2, 2.3, and 2.4 below.

The level of detail needed in the documentation should be sufficient to enable NRC to understand and determine the validity of all input data and calculation models used; to assess the sensitivity of the results to all key aspects of the analysis; and to audit any calculation. It is not necessary to submit all the documentation needed for such an NRC review, but its existence should be cited and it should be available in easily usable form. The guideline for adequate retained documentation is that an independent expert analyst should be able to reproduce any portion of the results of calculations in a straightforward, unambiguous manner. To the extent possible, the retained documentation should be organized along the lines identified in the areas of review.

A complete severe accident assessment requires analysis of external events. Previous guidance documents have discussed procedures for performing such analyses (NUREG/CR-2300 (Ref. 4) and NUREG/CR-2815 (Ref. 6)), and several full-scope PRAs and NRC's reviews of these PRAs have addressed external events. There is a technical basis for analyzing whether a given plant has significant vulnerabilities with respect to a given external initiator. Although IPE submittals are not presently required to address external events, it may be beneficial for utilities to be aware of such a future possibility, and they should retain information accordingly. Section 2.5 provides a discussion of future external-event analysis.

2.1 Front-End Submittal: Probability of Severe Accidents

The format and content of the front-end portion of the IPE submittal is addressed for the following key areas:

1. General Methodology
2. Information Assembly
3. Accident Sequence Delineation
4. System Analysis
5. Quantification Process
6. Front-End Results and Screening Process

Reporting guidelines for each of these key areas are detailed in Sections 2.1.1 through 2.1.6.

Table 2.1 Standard Table of Contents for utility submittal.

	Corresponding Section in This Report

Table 2.1 (Continued).

	Corresponding Section in This Report
3.4 Results and Screening Process	2.1.6
3.4.1 Application of Generic Letter Screening Criteria	
3.4.2 Vulnerability Screening	
3.4.3 Decay Heat Removal Evaluation	
3.4.4 USI and GSI Screening	
4. Back-End Analysis	
4.1 Plant Data and Plant Description	2.2.2.1
4.2 Plant Models and Methods for Physical Processes	2.2.2.2
4.3 Bins and Plant Damage States	2.2.2.3
4.4 Containment Failure Characterization	2.2.2.4
4.5 Containment Event Trees	2.2.2.5
4.6 Accident Progression and CET Quantification	2.2.2.6
4.7 Radionuclide Release Characterization	2.2.2.7
5. Utility Participation and Internal Review Team	2.4
5.1 IPE Program Organization	
5.2 Composition of Independent Review Team	
5.3 Areas of Review and Major Comments	
5.4 Resolution of Comments	
6. Plant Improvements and Unique Safety Features	2.3
7. Summary and Conclusions (including proposed resolution of USIs and GSIs)	

2.1.1 General Methodology

Reporting guidelines include a concise description of major tasks of the methodology employed and how these tasks interact with each other to generate the list of plant vulnerabilities. This includes such major tasks as event tree modeling, systems analysis, dependency treatment, quantification process, and vulnerability identification and treatment.

2.1.2 Information Assembly

Reporting guidelines include:

1. Plant layout and containment building information not contained in the Final Safety Analysis Report (FSAR).

2. A list of PRA studies or IPEs of this plant, or other similar plants, that the IPE team has reviewed along with a list of important insights derived from these reviews.

3. A concise description of plant documentation used such as the FSAR; system descriptions, procedures, and licensee event reports; and a concise discussion of the process used to confirm that the IPE represents the as-built, as-operated plant. The intent of such a confirmation is not to propose new design reverification efforts on the part of the licensees but to account for the impact of previous plant modifications or modifications conducted within the IPE framework.

4. A description of the walkthrough activity of the IPE team, including scope and team makeup.

2.1.3 Accident Sequence Delineation

Reporting guidelines include:

1. A list of all generic and plant-specific initiating events and groups of events considered (including internal flooding), their frequencies, and the rationale for the grouping used. Additionally, list the minimum success criteria for front-line systems that mitigate each initiating event or group of events, the bases for those criteria (e.g., expert judgment, realistic calculation, FSAR), and the consistency of the criteria with the as-built, as-operated plant. Refer to Reference 9 for additional insights on internal flooding.

2. All event trees (functional or systemic) developed or adapted from a reference plant for the initiating events or groups of initiating events, including a concise discussion of the assumptions and event heading dependencies considered.

3. If separate event trees are developed to support special event analysis (e.g., ATWS, station blackout, PWR reactor coolant pump seal loss-of-coolant accidents (LOCAs), interfacing-system LOCA, internal flooding), include the same information as in item 2 above.

4. The support system event trees, as applicable, including modifications if they have been adapted from the IDCOR reference plant or other applicable PRAs. A concise description of each of the support system states (or bins) found to be important and their effects on each of the front-line systems should be included.

5. An explanation of the method of grouping accident sequences into various "bins," "categories," or "plant damage states," including the unique bins considered and their physical meanings in terms of controlling factors such as initiating events, time of core melt, and performance of containment safety features.

6. A table summarizing the bins associated with the functional or systemic accident sequences that lead to core melt.

2.1.4 System Analysis

Reporting guidelines include:

1. A description and a simplified diagram of front-line and support systems considered in the IPE (e.g., appropriate line diagrams of electrical systems).

2. All fault tree diagrams should be retained by the utility and should be readily available upon request. The fault trees will be reviewed and audited on a case-by-case basis and need not be included as part of the IPE submittal.

3. The dependency matrix for all support systems and front-line systems (or functions) considered, including all functional interdependencies among the systems. This also includes dependencies caused by systems that are shared among multi-unit plants. Spatial or phenomenological dependencies that are scenario dependent should be discussed under Section 2.1.3, item 2. The discussion should describe how these dependencies were accommodated.

4. Differences between the subject plant and the reference plant if the dependency matrix is adapted from a reference plant. Identify modifications made to the matrix to reflect these differences.

5. Method used for determining unavailability of plant hardware, including a description of the unavailability consideration for standby and operating equipment and equipment in test and maintenance. Also, list any generic failure data used for equipment, equipment unavailability, or initiating events.

2.1.5 Quantification Process

Reporting guidelines include:

1. Types of common-cause failures considered in the analysis (both in the event tree sequences and in the system analysis), including the quantification process employed and sources of common-cause failure data used. Include a list of component groups subjected to common-cause failure analysis.

2. Internal flooding initiators such as overfilling of water tanks, hose and pipe ruptures, and pump seal leaks along with their frequencies and resulting damage to important plant equipment, including water intrusion. Include the result of the quantification of the flooding sequences that lead to core damage.

3. Types of human failures considered in the IPE, such as human failures in maintenance and operation and human failure to recover and mitigate accident progression.

4. List of human reliability data and time available for operator recovery actions considered, including the sources of these data. If the human errors are screened, include a list of errors considered and a list of "important errors," as well as the criteria for determining importance.

5. List of items for which plant-specific experience is used, including the method of generating failure data from such experiences (e.g., classical or Bayesian method). Include the rationale if plant-specific experience for initiating events and important items such as auxiliary feedwater and emergency core cooling system pumps, batteries, feed pumps, electrical buses, breakers, and diesel generators has not been used. (Generally, plants with several years of experience should use plant-specific experience for these types of items.) Also list any generic failure data used for equipment or initiating events.

6. Method by which accident sequences are quantified. If computer programs are used, identify the program and nature of calculations performed by using this program (e.g., cutset generation, sequence quantification, and sensitivity analysis).

2.1.6 Front-End Results and Screening Process

Reporting guidelines include:

1. For functional sequences, a description of how the screening criteria in Appendix 2 to the Generic Letter (Ref. 1) are used in the screening process. As an alternative, systemic sequences can be used provided the screening criteria given below are used to determine which potentially important systemic sequences and system failures (based on the procedure established in Ref. 4) that might lead to core damage or unusually poor containment performance should be reported to the NRC in the IPE submittal. It should be noted that, as with the functional screening criteria, these sequence criteria do not represent a threshold for vulnerability but only act as a reporting requirement. All numeric values given are "expected" (mean) values. The total number of unique sequences to be reported should be determined by the criteria listed below, or by the criteria in Appendix 2 to the Generic Letter, but in any case should not exceed the 100 most significant sequences. Sequences meeting more than one criterion should also be identified.

   a. Any systemic sequence that contributes 1E-7 or more per reactor year to core damage.

b. All systemic sequences within the upper 95 percent of the total core damage frequency.

c. All systemic sequences within the upper 95 percent of the total containment failure frequency.

d. Systemic sequences that contribute to a containment bypass frequency in excess of 1E-8 per reactor year.

e. Any systemic sequence that the utility determines from previous applicable PRAs or by utility engineering judgment to be an important contributor to core damage frequency or poor containment performance.

For systemic sequences, provide a description of how the above criteria are used in the screening process. For mixed sequences (part functional, part systemic), use systemic screening criteria. Because of overlap, sequences need only be reported once under any one of the criteria.

It should be noted that, for reporting purposes, all sequences (functional or systemic) should contain the initiating event (both the systems and containment responses), containment failure mode and timing, and estimated source term.

Analysts should be aware that it may not be prudent to terminate sequences arbitrarily just because they fall below the screening criteria, and therefore the screening criteria are to be used for reporting purposes only.

2. A list of sequences selected using the screening criteria, including a concise discussion of accident progression, specific assumptions, sensitive assumptions and parameters, essential equipment subjected to environmental conditions beyond the design bases and those conditions, and applicable human recovery actions.

3. A list of major contributors to those accident sequences selected using the screening criteria. Major contributions such as those from front-line systems or functions and support states, as well as contributions from unusually poor containment performance, are important for inclusion. Also include an estimate of total core damage frequency.

4. A thorough discussion of the evaluation of the decay heat removal function because the adequacy of the decay heat removal capability at the plant for preventing severe accident situations is to be resolved within this examination program. Plants without feed-and-bleed capability should particularly address the capability of the plant to recover from loss of all feedwater events (Refs. 13, 14, and 15). For purposes of the IPE, only power operation and hot standby need to be considered.

5. A list of any vulnerabilities identified by the review process, a concise discussion of the criteria used by the utility to define vulnerabilities, and the fundamental causes of each vulnerability. Vulnerabilities associated with the decay heat removal function should be specifically highlighted.

6. Identification of sequences that, but for low human error rates in recovery actions, would have been above the applicable core damage frequency screening criteria. Therefore, in addition to sequences reported under the screening criteria, any sequence that drops below the core damage frequency criteria because the frequency has been reduced by more than an order of magnitude by credit taken for human recovery actions should be discussed. Include information on the timing and complexity of postulated human actions. (The total number of sequences reported here should not exceed 50 of the most significant sequences.)

7. If applicable, include a discussion of other evaluations regarding the unresolved safety issues (USIs) and generic safety issues (GSIs) that have been assessed, including a discussion of the technical basis for resolutions proposed by the licensee for any USI or GSI. The following should be discussed:

    a. The ability of the methodology to identify vulnerabilities associated with the USI or GSI being addressed.

    b. The contribution of each USI or GSI to core damage frequency or unusually poor containment performance, including sources of uncertainty.

    c. The technical basis for resolving the issue.

    See Reference 16 for a listing and status of all USIs and GSIs.

2.2 Back-End Submittal: Containment Response

The IPE analysts must keep in mind the main objectives of performing the back-end study. The primary objective is to provide the utility with a framework for obtaining an understanding of and appreciation for containment failure modes, the impact of phenomena and plant features, and the impact of operator actions. The evaluation may also suggest areas for which additional training, formal procedures, or equipment modifications would improve the utility's ability to prevent or mitigate specific severe accidents. The second objective is to segregate out, over a broad spectrum of credible accidents, specific vulnerabilities associated with containment and containment mitigating systems. In some accident scenarios, specific vulnerabilities may be reduced or eliminated by enhancing procedures or improving mitigating system performance. By achieving these objectives, an appreciation of procedures, mitigating system performance, and mitigating system resources (e.g., electrical power, water, instrument air) will be achieved. These insights will allow for the evolution of an effective accident management program.

2.2.1 General Methodology

The general methodology for containment response has been described in Appendix 1 to the Generic Letter (Ref. 1). Although there is no unique way to perform the back-end analysis, Appendix A to this report provides additional insights and Appendix B provides useful reference material. Additional, potentially important material may be found in the Containment Loads Working Group Report (Ref. 17), the PRA Procedures Guide (Chapter 7 of Ref. 4), and draft NUREG-1150 (Ref. 18) and its supporting documents (Refs. 19 through 32).

On phenomenological matters, the status of the NRC position versus that of IDCOR was summarized in a series of so-called "issue" papers (Ref. 33). The utilities are expected to be cognizant of the methods and uncertainties reflected in those papers. Regarding the probabilistic treatment of phenomenological uncertainties, some additional material may be found in the Peer Review of Draft NUREG-1150 (Ref. 34). Appendix B contains a list of PRAs either performed or reviewed by the NRC. These documents, especially the more recent ones, may be useful in finding suitably similar plants and sequences to aid back-end analyses. The NRC comments on industry-sponsored PRAs should be kept in mind if those results are used.

2.2.2 Specific Guidelines

In order to facilitate and ensure a high-quality review process, each submittal should be organized in major sections as follows (see Table 2.1):

1. Plant Data and Plant Description
2. Plant Models and Methods for Physical Processes
3. Bins and Plant Damage States (Interface with Front-End Assessment)
4. Containment Failure Characterization
5. Containment Event Trees
6. Accident Progression and CET Quantification
7. Radionuclide Release Characterization

2.2.2.1 Plant Data and Plant Description

Identify and highlight component, system, and structure data that may be of significance in assessing severe accident progressions. Additional consideration should be given to equipment whose operability is desired during exposure to harsh environments. Describe systems such as fan coolers or sprays that are important to operation during a severe accident. This description should extend to the reactor building or auxiliary building if appropriate. The utility has the option of submitting a concise set of the plant data that is relevant to severe accident phenomenology or an identification of those containment features that are unique to the facility in question relative to the similar plant that was the subject of previous PRAs such as those for NUREG-1150 (Ref. 18). In addition to the appropriate narrative explanations and sketches, this information should be summarized in tabular form.

The assessment of the "significance" of such unique features may, of course, be judgmental and based upon the understanding of severe accident phenomena and associated containment challenges developed through the IPE. For example, debris bed coolability depends heavily on such plant features as available spread area within the cavity and water availability in the cavity. Both aspects are highly individualized even among plants of the same type; thus, an accurate but straightforward representation of such plant features would be needed.

The process of providing sufficient plant data gets more complicated when considering mechanisms that are incompletely understood. For example, it is agreed that phenomena associated with high-pressure melt ejection depend heavily on the characterization of the vessel's lower head, the sizes of the flow paths within and out of the reactor cavity, and the lower subcompartment geometry, although actual relationships to resulting containment loads are lacking. Similarly, the potential for non-uniform distribution of combustible gases in the containment air space is clearly related to geometry and to location, composition, and intensity of release; however, little basis exists for judging which are important features and the extent of their impact on mixing. It is requested, therefore, that accurate but simple representations of containment geometry be made in this section in as complete a fashion as possible so as to cover the needs in the two cases mentioned above and possibly other situations as they might arise in the submittal's treatment of phenomenology. While blueprints are not necessary, drawings that accurately display the location and rough dimensions of components, systems, and structures that are important for accident progression assessment should be included.

2.2.2.2 Plant Models and Methods for Physical Processes

Provide concise documentation of all analytical models, including selection of empirical factors and data inputs, used in the accident progression analysis. Well-known codes and published models, or even widely accepted results on particular aspects of the phenomenology, may be incorporated simply by reference. To the extent that accepted results can be used, the utility can gain the insights about physical processes without the effort of de novo analysis and without extra review by the staff. For example, if the utility chooses to use CORCON for core-concrete interactions, it can do so provided reference is given to the specific modification to CORCON that is used. General assumptions used in the modeling of phenomenology are just as important as the models themselves and therefore should be fully described. Organization should be such that all particular results quoted in subsequent sections can be referred conveniently to respective analytical models of this section. Clearly, fully integrated analytical tools may not be necessary; however, it is important that the composing of overall accident behavior from separate effects analyses be clearly delineated.

2.2.2.3 Bins and Plant Damage States

As in standard methodology, the coupling of the front-end analysis to the back-end is through the binning of the multitude of front-end sequences into a few groups of damage states with similar back-end characteristics. It is important that the bins be justified on the basis of such factors as timing of important events or operability of key features. Also, the state of the various systems and components, as deduced from the detailed front-end analysis, should be accurately translated into the back-end plant damage states considered. The impact of severe accident phenomena on the operability of such systems and components must be reflected where appropriate.

Accordingly, this section, in a manner consistent with the binning guidelines of Section 2.1.3 (items 5 and 6), should concisely cover or reference the methodology and results of binning, as well as the actual procedures employed. Further, all front-to-back-end sequence interfaces (i.e., reactor coolant system and containment thermal-hydraulic conditions, containment mitigation system availability, support system availability, human factor assumptions) need to be concisely documented, and the adopted binning needs to be justified. Care should be taken to properly bin sequences that will progress under different thermal-hydraulic conditions; for example, high reactor coolant system pressure versus low reactor coolant system pressure or different timing--slowly developing or fast. Binning should facilitate further evaluation of potential preventive or mitigative measures.

Recent studies, such as NUREG-1150 (Ref. 18), have stressed the importance of mission times, inventory control (of such resources as instrument air or battery power), and dual usage (e.g., when the condensate storage tank supplies water for both vessel injection and containment sprays, early injection may deplete the water so that it is not available for sprays). Therefore, for the screened sequences, it is important that the impact of mission times, inventory control, and dual usage be discussed with respect to the progression of the accidents, the estimated frequencies, and the binning process.

2.2.2.4 Containment Failure Characterization

This section should provide comparisons with structural calculations for other plants of similar design performed (or results of structural calculations if the licensee has chosen to perform such analyses) to assess containment strength and the magnitude of various loads necessary to fail containment, e.g., static pressure, localized heat loads, and localized dynamic pressures. A sample list of potential containment failure modes and mechanisms is provided in Table 2.2; these have been considered in Reference 18. Other failure mechanisms may be appropriate for specific designs. Some of the modes in Table 2.2 are more important for some containment designs than for others. If the analysts choose to incorporate results obtained previously for other containments, it is important to provide a concise rationale of their applicability. The vulnerability of containment penetrations to thermal attack is discussed in Reference 35. The licensee submittal should include an assessment of the penetration elastomer seal materials and their response to prolonged high temperatures. Particular attention should be paid to seals in areas where standing hydrogen flames are possible.

In each case, potential failure locations should be identified together with estimated failure sizes.

Finally, an assessment of failure size and location should be made for any other structures within which radionuclide transport and retention will be considered (e.g., as-built vent piping and the reactor building in BWRs).

2.2.2.5 Containment Event Trees

The first containment event tree (CET) nodal decision point should determine the likelihood of whether the containment is isolated, bypassed, intact, or failed (i.e., a branch point split fraction). For those IPEs that have found this to be impractical and have treated containment isolation elsewhere in the analyses, the process should be described. In either case, the analyses should address the five areas identified in the Generic Letter, i.e., (1) the pathways that could significantly contribute to containment isolation failure, (2) the signals required to automatically isolate the penetration, (3) the potential for generating the signals for all initiating events, (4) the examination of the testing and maintenance procedures, and (5) the quantification of each containment isolation failure mode (including common-mode failure).

Table 2.2 Potential containment failure modes and mechanisms.

Direct bypass

Failure to isolate

Vapor explosions

      Missile generation
      Quasi-static pressure rise

Overpressurization

      Steam
      Noncondensible gases

Combustion processes (hydrogen, carbon monoxide, methane)

      Blast
      Quasi-static pressure rise

Core-concrete interaction

      Basemat penetration
      Structural failure and tearout of penetrations

Blowdown forces

      Vessel thrust force

Meltthrough

      Direct contact of containment shell with fuel debris

Thermal attack of containment penetrations

It is important to note that this section is closely coupled to the following section (2.2.2.6), "Accident Progression and CET Quantification." Not only does Section 2.2.2.6 quantify the split fractions for the CETs, but, depending on the results of the accident progression analysis, it could dictate the structure of the CETs themselves.

All accident sequences (represented now by plant damage states or bins) that meet the screening criteria should be represented by CETs according to standard practice. Helpful guides and standard practice concerning the structure and methods of analysis of CETs can be found in a number of back-end PRAs as listed in Appendix B but subject to the comments reported in the NRC review reports.

2.2.2.6 Accident Progression and CET Quantification

The submittal should present a characterization of containment performance for each of the CET end-states based on assessment of loads. Significant loads are those with the potential to challenge containment integrity. In this interpretation, the containment boundary should be taken to include any interface with a more or less direct access to the outside (e.g., primary-to-secondary pressure boundary, drywell shell in Mark I). Each predicted load should be adequately supported by reference to either:

1. A particular model presented in Section 2.2.2.2 or
2. A previously published (i.e., referenceable) analysis.

In the latter case, applicability would be established through comparison of geometry and thermal-hydraulic conditions. Appendix 1 to the IPE Generic Letter (Ref. 1) provides guidance for assessing containment loads. Additional insights can be found in Appendix A to this document. NRC-sponsored calculations of containment loads that take into account certain phenomenological and containment loading issues can be found in the supporting documentation for Reference 18 (Refs. 19 through 32). In any event, selected pressure and temperature histories for representative CET end-states should be displayed graphically for the containment compartments and other building compartments of interest.

On the basis of the above and any additional pertinent analyses, this section continues with the quantification of the CETs. In the quantification of the CET, human intervention would be based on existing emergency operating procedures (EOPs) and assessed against standards for human performance or those planned for near-term implementation. If EOPs are used in controlling or ameliorating the outcome of the accident, the submittal should describe the operational status of these EOPs and verify that the required amount of training has been (or will be) performed.

Documentation should be provided to support the availability and survivability of systems and components with potentially significant impact on the CET or the radionuclide release. The equipment environment should be assessed with the same temperature, pressure, humidity, and radiation environment predicted as part of the accident progression analysis. The utility should pay particular attention to equipment vulnerability and survivability. If containment sprays, for example, are operating to remove heat and wash out radionuclides, the utility should assess the capability of the system to perform its function for the allotted time under the expected environmental conditions. Time is an important consideration, especially for accident sequences that do not fail the containment early in the sequence. Additional details may be required to justify time and component reliability during such harsh environmental conditions. Reference 36 provides additional information and insights into potential risk-significant equipment qualification issues.

It should be noted here, however, that the intent is not to extend 10 CFR 50.49 equipment qualification criteria to beyond design basis (severe accident) conditions. The intent is to emphasize the need for sound engineering judgment, corrective action where appropriate, and consideration of equipment survivability within the evolution and framework of the IPE and subsequent accident management program.

A description should be given of information used in determining the conditional probability that the containment is not isolated, given a core melt accident, including capability, testing, trip signals, overrides, diagnostics, and, of course, experience. (This is the so-called "beta failure mode" for containments as used in PRAs.) In addition to the conditional probability, a description of the size and characterization of the isolation failure should be included.

A description should be given of the assessment of accident sequences that bypass the containment (interfacing-system LOCA). Reference 3 discusses the plant features found to be important.

Finally, this section should make clear the methods employed for handling phenomenological uncertainties in this quantification. The staff recognizes that there are significant unresolved phenomenological uncertainties associated with the quantification of containment event trees. The purpose for considering uncertainties is to avoid the masking of potential vulnerabilities due to technically unsupportable assumptions regarding the likelihood of certain phenomena. The uncertainty consideration may be either quantitative or qualitative. The submittal should describe the process in sufficient detail so that the reviewer may have confidence that phenomenological and other uncertainties have been properly accounted for in the identification of candidate plant improvements. (See Steps 5 and 8 of Appendix A for an approach to addressing this part of the analysis.)

2.2.2.7 Radionuclide Release Characterization

Quantification of the CETs will produce estimates of the probability and mode of containment failure for the various plant damage states identified. By combining the frequencies of the plant damage states with the probabilities of the various failure modes, the frequencies of containment failure or bypass can be determined. If a sequence is found to have a core damage frequency that exceeds the screening criteria, the magnitude of the radionuclide release should be estimated. Determination of the source term should require significantly less resources if the analyst chooses to use existing calculations for similar plants and sequences in lieu of de novo calculations.

This may be done by selection of source terms for similar sequences that have been identified for a similar plant or by code calculation. References 37 and 38 contain calculations that provide source term information. Whatever approach is used, concise documentation should be provided as to how release characteristics were assigned. If a code is used, it should be referenced and the input assumptions provided.

If a large number of source term calculations are combined into a set of release categories, the rationale for the process should be provided. If sequences are binned prior to calculating a single source term for a representative sequence in the bin, the rationale for the binning process and for the selection of the representative sequence should be provided.

The staff encourages assessment of accident management issues concurrent with the performance of the IPE since the results of the IPE will be a major source of information for use by the utility in developing its accident management program. For instance, the inventory of radionuclides residing in areas to which personnel may need access (e.g., reactor building, auxiliary building) may be identified in the IPE and used by the utility to determine the feasibility of attempting to restore components in those areas as part of accident management.

Containment failure mode and timing are of primary importance. However, radio-active material release and transport through the reactor coolant system, the containment, and auxiliary buildings must also be considered. Only through con-sideration of where radioactive material might be at any given time in a sequence can such issues as operator actions (e.g., if operators perform tasks in the radiation environment that might exist) or equipment performance (e.g., the aerosol and radiation level that equipment will be expected to withstand) be fully assessed.

The section should conclude with the ranking of release categories on the basis of both their conditional and total (i.e., including front-end results) probabilities.

2.3 Submittal of Specific Safety Features and Potential Plant Improvements

On the basis of the understanding developed through the IPE, the utility should develop and document in this section a list of any specific safety features that are believed to be unique and/or important to the facility. Among the family of such features would be those features that resulted in significantly lowering what are considered to be high-frequency core melt sequences or acci-dent progressions in contemporary PRAs for similar plants.

The utility should document any worthwhile strategies to further prevent or mitigate the detrimental effects of severe accidents that were developed as part of the IPE process and for which credit has been taken in the analysis. For the vulnerabilities from the functional or systemic sequences, the utility should identify potential improvements, if any, including equipment changes as well as changes in maintenance, operating and emergency procedures, surveillance, and training programs that have already been implemented or have been selected for implementation. Include a discussion of the anticipated benefits in terms of the vulnerabilities addressed. Downside considerations should also be addressed. If all the potential improvements have been dropped from further consideration because of the high cost, it is important to discuss how less expensive alternatives were sought. Not all strategies that were considered during the IPE process need to be included in the final report. If a strategy has been selected for implementation to address a vulnerability, for example, only that strategy need be described. The submittal should provide enough documentation so that the reviewer can be confident that a reasonable effort to address each identified vulnerability has been performed, whether or not a fix has been implemented. Describe the rationale by which potential options were selected for implementation. Provide, in tabular form, which options have been scheduled for implementation and the respective timing of implementation.

For those potential improvements that would only be under consideration because of the unresolved generic phenomenological issues in the NRC Containment Performance Improvements Program (for example, an improvement that would only be justified if direct containment heating caused early containment failure), the staff has made it clear in the Generic Letter that the industry will not be placed in a position of having to implement improvements before all containment performance decisions have been made. However, consistent with the IPE Generic Letter, the submittal should "...develop strategies to minimize the challenges and the consequences such severe accident phenomena may pose to the containment integrity and to recognize the role of mitigation systems while awaiting their generic resolution."

2.4 IPE Utility Team and Internal Review

The basis for the request in the Generic Letter (Ref. 1) for involvement of utility staff in the IPE review is the belief that the maximum benefit from the performance of an IPE would be realized if the utility's staff were involved in all aspects of the examination and that involvement would facilitate integration of the knowledge gained from the examination into operating procedures and training programs. Thus the submittal should describe utility staff participation and the extent to which the utility staff was involved in all aspects of the IPE program.

The Generic Letter requests that each utility conduct "...an independent in-house review to ensure the accuracy of the documentation packages and to validate both the IPE process and its results." The submittal should contain, as a minimum, a description of the internal review performed, the results of the review team's evaluation, and a list of the review team members.

The purpose of the in-house review is twofold. First is the importance of having utility personnel cognizant of the IPE. The maximum benefit to the utility would occur if the combination of persons involved in the original analysis and in-house review, taken as a group, provides both a cadre of utility personnel to facilitate the continued use of the results and the expertise in the methods to ensure that the techniques have been correctly applied.

The second purpose of the in-house review is to provide quality control and quality assurance to the IPE process. Independence of the review team is desirable because it reflects a quality control and quality assurance attitude. In situations where it is necessary to use a reviewer who has not been totally removed from the plant-specific IPE process, the utility should have confidence that the reviewer can be objective and capable of providing critical review. The utility may wish to solicit outside reviewers from an adjacent unit in order to achieve a certain degree of objectivity in the review process. In any case, the staff expects that all utilities have in-house the most expert knowledge of their own plants, system configurations, and operating practices and procedures.

2.5 Consideration of External Events

The IPE Generic Letter (Ref. 1) states that examination of external events will proceed separately and on a later schedule from that of the internal events. Because of this, no reporting for external-event analysis is required at this time. However, utilities may choose to submit their examinations of external events at this time as part of the IPE program. The external-event analyses submitted will be evaluated during the IPE review process on a case-by-case basis.

It may be prudent for the utilities to properly retain documents and plant-specific data relevant to external events such that they can be readily retrieved for future external-event analyses. This minimizes the need for a second performance of similar tasks and allows maximum utilization of the internal-event analysis, models, and data. Early consideration of some special aspects of external events such as spacial dependencies will also prove to be beneficial by extending the usefulness of the internal-event fault trees when external-event analyses are conducted.

REFERENCES

1. NRC letter to All Licensees Holding Operating Licenses and Construction Permits for Nuclear Power Reactor Facilities, "Individual Plant Examination for Severe Accident Vulnerabilities - 10 CFR §50.54(f)," Generic Letter No. 88-20, dated November 23, 1988.

2. W. T. Pratt, "Severe Accident Insights Report," Brookhaven National Laboratory, NUREG/CR-5132, BNL-NUREG-52143, April 1988.

3. Brookhaven National Laboratory, "Assessment of Severe Accident Prevention and Mitigation Features," NUREG/CR-4920, Vols. 1-5, BNL-NUREG-52070, July 1988.

4. J. W. Hickman, "PRA Procedures Guide: A Guide to the Performance of Probabilistic Risk Assessments for Nuclear Power Plants," American Nuclear Society and Institute of Electrical and Electronic Engineers, NUREG/CR-2300, Vols. 1 and 2, January 1983.

5. A. El-Bassioni et al., "PRA Review Manual," Brookhaven National Laboratory, NUREG/CR-3485, BNL-NUREG-51710, September 1985.

6. M. McCann et al., "Probabilistic Safety Analysis Procedures Guide," Brookhaven National Laboratory, Revision 1 to NUREG/CR-2815, Vols. 1 and 2, August 1985.

7. Industry Degraded Core Rulemaking (IDCOR) Program, "Individual Plant Evaluation Methodology for LWRs," April 1987.

8. Letter from A. Thadani, NRC, to W. Rasin, NUMARC, "Staff Evaluation of IDCOR IPEMs," dated November 22, 1988.

9. D. Thatcher, "Evaluation of Systems Interactions in Nuclear Power Plants," NUREG-1174, May 1989.

10. A. D. Swain III, "Accident Sequence Evaluation Program--Human Reliability Analysis Procedure," Sandia National Laboratories, NUREG/CR-4772, SAND86-1996, February 1987.

11. L. M. Weston et al., "Recovery Actions in PRA for the Risk Methods Integration and Evaluation Program," Sandia National Laboratories, NUREG/CR-4834, Vols. 1 and 2, SAND87-0179, June 1987.

12. L. N. Haney et al., "Comparison and Application of Quantitative Human Reliability Analysis Methods for Risk Method Integration and Evaluation Program," Idaho National Engineering Laboratory, NUREG/CR-4835, EGG-2485, January 1989.

13. L. Marsh and C. Liang, "Evaluation of the Need for a Rapid Depressurization Capability for Combustion Engineering Plants," NUREG-1044, December 1984.

14. USNRC, "Power-Operated Relief Valves for Combustion Engineering Plants," SECY-84-134, dated March 23, 1984.

15. D. R. Gallup et al., "Cost/Benefit Analysis of Adding a Feed and Bleed Capability to Combustion Engineering Pressurized Water Reactors," Sandia Laboratories, NUREG/CR-3421, October 1983.

16. R. Emrit et al., "A Prioritization of Generic Safety Issues," NUREG-0933, Supplement 8, November 1988.

17. U.S. Nuclear Regulatory Commission (USNRC), "Estimates of Early Containment Loads from Core Melt Accidents," NUREG-1079, Draft Report for Comment, December 1985.

18. USNRC, "Severe Accident Risks: An Assessment for Five U.S. Nuclear Power Plants," NUREG-1150, Vols. 1 and 2, Second Draft for Peer Review, June 1989.

19. D. M. Ericson, Jr., (Ed.) et al., "Analysis of Core Damage Frequency: Methodology Guidelines," Sandia National Laboratories, NUREG/CR-4550, Vol. 1, Rev. 1, SAND86-2084, to be published.*

20. T. A. Wheeler et al., "Analysis of Core Damage Frequency from Internal Events: Expert Judgment Elicitation," Sandia National Laboratories, NUREG/CR-4550, Vol. 2, SAND86-2084, April 1989.

21. R. C. Bertucio and J. A. Julius, "Analysis of Core Damage Frequency: Surry Unit 1," Sandia National Laboratories, NUREG/CR-4550, Vol. 3, Rev. 1, SAND86-2084, to be published.*

22. A. M. Kolaczkowski et al., "Analysis of Core Damage Frequency: Peach Bottom Unit 2," Sandia National Laboratories, NUREG/CR-4550, Vol. 4, Rev. 1, SAND86-2084, to be published.*

23. R. C. Bertucio and S. R. Brown, "Analysis of Core Damage Frequency: Sequoyah Unit 1," Sandia National Laboratories, NUREG/CR-4550, Vol. 5, Rev. 1, SAND86-2084, to be published.*

24. M. T. Drouin et al., "Analysis of Core Damage Frequency: Grand Gulf Unit 1," Sandia National Laboratories, NUREG/CR-4550, Vol. 6, Rev. 1, SAND86-2084, to be published.*

25. M. B. Sattison and K. W. Hall, "Analysis of Core Damage Frequency: Zion Unit 1," Idaho National Engineering Laboratory, NUREG/CR-4550, Vol. 7, Rev. 1, EGG-2554, to be published.*

26. E. D. Gorham-Bergeron et al., "Evaluation of Severe Accident Risks: Methodology for the Accident Progression, Source Term, Consequence, Risk

---

*Available in the NRC Public Document Room, 2120 L Street NW., Washington, DC.

Integration, and Uncertainty Analyses," Sandia National Laboratories, NUREG/CR-4551, Vol. 1, Draft Revision 1, SAND86-1309, to be published.*

27. F. T. Harper et al., "Evaluation of Severe Accident Risks: Quantification of Major Input Parameters," Sandia National Laboratories, NUREG/CR-4551, Vol. 2, Draft Revision 1, SAND86-1309, to be published.*

28. R. J. Breeding et al., "Evaluation of Severe Accident Risks: Surry Unit 1," Sandia National Laboratories, NUREG/CR-4551, Vol. 3, Draft Revision 1, SAND86-1309, to be published.*

29. A. C. Payne, Jr., et al., "Evaluation of Severe Accident Risks: Peach Bottom Unit 2," Sandia National Laboratories, NUREG/CR-4551, Vol. 4, Draft Revision 1, SAND86-1309, to be published.*

30. J. J. Gregory et al., "Evaluation of Severe Accident Risks: Sequoyah Unit 1," Sandia National Laboratories, NUREG/CR-4551, Vol. 5, Draft Revision 1, SAND86-1309, to be published.*

31. T. D. Brown et al., "Evaluation of Severe Accident Risks: Grand Gulf Unit 1," Sandia National Laboratories, NUREG/CR-4551, Vol. 6, Draft Revision 1, SAND86-1309, to be published.*

32. C. K. Park et al., "Evaluation of Severe Accident Risks: Zion Unit 1," Brookhaven National Laboratory, NUREG/CR-4551, Vol. 7, Draft Revision 1, BNL-NUREG-52029, to be published.*

33. T. Speis, USNRC, letters to A. Buhl, International Technology, dated September 22, 1986, and November 26, 1986.

34. W. E. Kastenberg et al., "Findings of the Peer Review Panel on the Draft Reactor Risk Reference Document, NUREG-1150," Lawrence Livermore National Laboratory, NUREG/CR-5113, UCID-21346, May 1988.

35. USNRC, "Containment Performance Working Group Report," NUREG-1037, Draft Report for Comment, May 1985.

36. L. D. Bustard et al., "EQ Risk Scoping Study," Sandia National Laboratories, NUREG/CR-5313, SAND88-3330, January 1989.

37. J. A. Gieseke et al., "Radionuclide Release Under Specific LWR Accident Conditions--PWR Large, Dry Containment Design (Surry Plant Recalculations)," Battelle Columbus Laboratories, BMI-2104, Vol. V, Draft, July 1984.

38. R. S. Denning et al., "Report on Radionuclide Release Calculations for Selected Severe Accident Scenarios," Battelle Columbus Laboratories, NUREG/CR-4624, Vols. 1-5, BMI-2139, July 1986.

---

*Available in the NRC Public Document Room, 2120 L Street NW., Washington, DC.

APPENDIX A

APPROACH TO BACK-END PORTION OF IPE

Introduction

Section 2.2 provides general guidance on the reporting of the back-end portion of the plant examination. This appendix provides a more specific approach on how the back-end portion of the examination could be performed. It should be noted that there is no unique way to perform this portion of the plant examination. The approach used may vary among different plant types and among analysts. This appendix provides information based on approaches that have been used in previous probabilistic risk assessments (PRAs). Some of these PRAs were sponsored by the NRC; others were sponsored by industry and reviewed by the NRC staff. As such, the information should be useful to utilities when performing their IPEs but should not be interpreted as a set of comprehensive requirements for performing a plant examination. The information provided in this appendix is based on the studies referenced in Appendix B.

The series of steps described below are intended to organize the activities that will be needed when performing a plant examination. These steps are similar to the subtasks identified in NUREG/CR-2300 (Ref. A.1). Although these steps are organized sequentially, in practice there will be considerable interaction among the tasks performed under each step. The organization of the tasks (or steps) is left to the individual analyst; however, the staff does expect that all the tasks identified in each of the steps in this appendix will be addressed in some form as part of an IPE.

Step 1 - Plant Familiarization

This step is described in general terms in Section 2.2.2.1 of this report. In this approach, the plant data would be displayed in three basic forms: tabular data, a descriptive narrative of key mitigative systems, and drawings of key civil structures and hardware. Systems such as containment spray would be described under Section 2.1.2 (Information Assembly), and references to front-end descriptions would be given. A test of whether sufficient descriptive material has been consolidated is that a reviewer should be able to reconstruct the sequences that are reported.

An example of important data for a PWR with a large, dry containment that might be provided in tabular form is given in Table A.1. Drawings included for the back-end submittal are basically those that the utility analysts found helpful in making their final assessment. Table A.2 provides a list of drawings for a PWR. (A parallel level of detail would be reasonable for BWRs.)

The narrative portion of the supporting information compiled and retained by the licensee might include a summary of operating experience and testing of mitigative systems. Containment isolation procedures and assurances should be included. In a general way, the narrative should fill in the gaps of information that would otherwise be incomplete if only the drawings and tabular data were provided.

The capability of key systems to perform their functions in severe accidents would be included. This discussion would address survivability under the

Table A.1 Examples of data to be assembled in tabular form
for back-end assessment.

1. Reactor Core Vessel and Primary System

A. Core and Vessel Data

Core full power, mass of UO_2 in core, mass of Zr in core; mass of Zr in cladding surrounding the fuel; mass of steel in core structures broken down into logical categories (upper plenum structures, core support plate, etc.); mass of bottom head; bottom head diameter and thickness; fuel enrichment; mass of control rod constituents.

B. Primary System Data

Total water inventory under normal full-power conditions; total water and steam volumes under normal full-power operation; type, number, and model of steam generators; total flow rate under normal full-power conditions; PORV capacities, safety valve capacities, and settings; normal reactor coolant system temperature, pressure, and enthalpy.

C. Accumulator Tanks

Total mass of water, inventory temperature, initiating pressure.

2. Containment System

A. Containment Structure

Containment type (steel, concrete, prestressed/posttensioned, re-enforced, etc.); type and chemical composition of concrete used in the basemat, including the weight fraction of free H_2O and bound H_2O; free volume; design pressure; normal (full-power) pressure; normal (full-power) temperature; area of reactor cavity floor;, containment liner thickness, wall thickness at key locations, and basemat thickness.

B. Containment Mitigation Systems--Sprays

Number of injection pumps, total design flow rate, containment setpoint for spray initiation, spray additives (if any).

C. Containment Mitigation Systems--Fan Coolers

Capacity, number of fans, flow rate per fan, primary inlet temperature.

D. Interior Structural Heat Sinks

Table A.1 (Continued).

3. ECCS and Other Water Injection/Recirculation Systems

   For all the systems listed below, total flow rates, number of pumps, and pressure setpoints:

   A. Volume/chemistry control charging pumps
   B. High-pressure injection
   C. Low-pressure injection
   D. Residual heat removal (RHR)
   E. Upper head injection (ice condenser containments)

   For refueling water storage tank:

   Total mass of water and initial (normal) temperature range.

4. Auxiliary Building (Reactor Building for BWRs)

   Data similar to above for all systems components and structures used in the IPE assessment to mitigate the consequences of a severe accident.

Table A.2 Examples of drawings to be provided for back-end assessment.

Drawings of the primary system--including detailed drawings of a typical vessel lower head instrument tube penetration; the vessel support plate region; and a steam generator. Drawings should be to scale unless clearly noted. The primary system drawings should show elevations and the pressurizer pressure relief and surge tank.

Drawings of the reactor cavity area--including the following information: cavity floor area; cavity sump; liner location and basemat thickness; cavity pressurization relief paths such as around the vessel, through the personnel access (if any), and through the instrumentation tube pathway; and vessel lower head location. The cavity elevation drawing should indicate the level of water in the cavity assuming all the primary system and RWST water has been injected into the containment (and failed vessel). Drawings should be to scale unless clearly noted and should contain sufficient dimensions for independent analysis.

Drawings of the containment building--both elevation and plan views with the following items highlighted:

- Location of sprays
- Location of fan cooling system, including ductwork
- Key structural features such as crane wall
- Location of primary system components/secondary system components/ accumulators/surge tank
- Location of containment sump systems
- Location of key penetrations
- Relation of containment building to auxiliary building
- Location of components and piping for ECCS and RHR systems
- Specific indication of any confined spaces that might accumulate combustible gases
- For ice condenser containments, additional items should include upper and lower compartments, ice chests, plenum areas, and air-return fans. Also location of hydrogen control devices (igniters).

Drawings of Auxiliary Building--showing relation to control room, containment building, emergency diesel building(s), and turbine building if such buildings are part of the flowpaths to the environment. Include the various routes for the release of radionuclides and noncondensible gases to the outside environment should steam generator tubes fail during a severe accident.

Concise discussion or simplified drawings of the primary system, the secondary system, the ECCS systems (injection and recirculation modes), the RHR system, the containment spray system, the fan cooling system, and the volume/chemical control system.

pressure, temperature, radiation, debris, and steam conditions expected during a severe accident. For example, what are the effects of debris aerosols and particulates on the operation of the sprays (in recirculation mode) and the fan coolers? The intent, however, is not to extend 10 CFR 50.49 equipment qualification criteria to beyond design basis conditions, but to use sound engineering judgment, as appropriate. Reference A.2 contains useful insights.

A discussion of operator actions that are taken to engage or maintain any of the above systems would be included. A list would be prepared of all manual operations of mitigative systems.

Step 2 - Sequence Grouping

This step is described in Section 2.2.2.3. The purpose of this step is to consolidate the large number of accidents that lead to core damage into a smaller number of plant damage states. This involves binning accident sequences into plant damage states that have approximately similar characteristics. The intent is that all accidents within a particular plant damage state can be treated as a group for the purpose of assessing accident progression, containment response, and fission product release. Those plant features that influence accident progression after core damage may vary between plant types. There is no one unique way to perform this analytical task. However, Table A.3 provides an example of a binning scheme that has been used for several PRAs for PWRs with large volume containments.

Once a binning scheme has been developed, the accidents defined in the front-end portion of the examination can be readily allocated to the appropriate plant damage state. A binning scheme such as the one shown in Table A.3 can generate between five and ten plant damage states for a typical PRA. Isolation failure is included in the binning scheme in Table A.3 and therefore the system portion of the plant examination must identify those core melt accidents that also have isolation failure as an initiating event so that they can be appropriately binned. Alternatively, isolation failure could have been dealt with as the first question in the containment event trees (refer to Step 4 below). This latter approach was the one suggested in Appendix 1 to the Generic Letter. Both approaches should lead to the same result.

After all the accident sequences are allocated to appropriate bins, one or more sequences would be selected to represent each plant damage state bin. Usually the accident sequence with the highest frequency is used to represent the bin, although other criteria may also be important. Whatever approach is used in the IPE, it is important to describe the process used to select the representative sequence. This is an important step in the examination process because these representative sequences will be used to quantify the containment event trees (CETs) (Step 7).

Step 3 - Determination of Containment Failure Modes

This step is described in Section 2.2.2.4 of this report. The first task is to identify a list of potential containment failure modes. Table A.4 provides a list of potential failure modes for five containment types that were identified by previous studies. The importance of these modes to other plants will be determined as part of the plant examinations. In addition, some plants may have failure modes that are not given in Table A.4. The information given in

Table A.3 Example of plant damage state bin characteristics.

Initiating Event	ECC Operation		Sprays Fail Operate	Fan Coolers Fail Operate
	Fails in Injection	Fails in Recirculation		
Small LOCA • • •	x		x	x
Intermediate LOCA • • •		x	x	x
Transients • • •	x		x	x
Interfacing-System LOCA • • •	x		n/a n/a	n/a n/a
Containment Isolation Failure • • •				

A-8

Table A.4 should therefore be interpreted only as a starting point for this step in the examination process and not as a comprehensive list or ranking of containment failures modes.

Another task in this step is to determine the structural capacity of the containment. Section 2.2.2.4 indicates that this task may be performed by carrying out plant-specific calculations or by using calculations that have been performed in the past for similar plant designs. Appendix B provides an extensive list of previous studies that could be used in the individual plant examinations. In many cases, these studies provide the necessary perspective to understand the progression of severe accidents and can be used to develop a hierarchy of containment failure modes and timing. If a utility chooses to use calculations that were performed in support of an industry PRA (Table B.2), they should also take into account the findings of the NRC staff's review of the PRA (Table B.3). If reference to the results of calculations that were performed in support of PRAs are not available to the NRC staff, such calculations should be made available if requested by the staff.

Step 4 - Develop Containment Event Trees

This step is addressed in Section 2.2.2.5 of this report. The most common approach to organizing the containment analysis portion of a PRA is to use a containment event tree (CET). Thus, for each plant damage state identified in Step 2, a containment event tree should be developed. Again there is no unique approach to CET development. CETs vary from the extremely simple trees developed for the IPEM (Ref. A.4) to the extremely complex trees used in draft NUREG-1150 (Ref. A.5). The staff determined (Ref. A.6) that the simplified IPEM event trees were too narrowly focused and that they were therefore unacceptable. However, the staff does not believe that a utility has to develop CETs as complex as those used in NUREG-1150 to perform an IPE.

A CET should provide sufficient nodal questions such that the important events that impact containment performance can be addressed and quantified (Step 7). Thus, as a minimum, all the pertinent containment failure modes identified in Step 3 would be included in the CET. Chapter 7 of NUREG/CR-2300 (Ref. A.1) provides a description of the development of CETs that is still largely applicable today. CETs that have been reviewed by the NRC staff and contractors are contained in Appendix B. The analyst should refer to these review documents to determine the staff's evaluation of the approach.

CETs are developed to describe the progression of an accident sequence. It is therefore convenient to set up the CETs as a series of time sequences. The following four time sequences are normally considered in one form or another in most CETs:

Time Period 1 - Events before core melt.
Time Period 2 - Events related to in-vessel phenomena.
Time Period 3 - Events related to out-of-vessel phenomena after vessel failure.
Time Period 4 - Events related to ex-vessel core debris disposition and coolability.

Table A.4 Potential containment failure modes for existing
plants identified by previous studies.

Potential Failure Modes	Large Volume	Ice Con- denser	Mark I	Mark II	Mark III
Containment Bypass					
• Interfacing-system loss-of-coolant accident	Yes[1]	Yes[1]	Yes[1]	Yes[1]	Yes[1]
• Failure to isolate containment	Yes[1]	Yes[1]	Yes[1]	Yes[1]	Yes[1]
• Steam generator tube rupture	Yes[1]	Yes[1]	N/A	N/A	N/A
Early Containment Failures					
• Overpressurization with high temperatures					
- due to noncondensible gases and steam	Yes[2]	Yes[2]	Yes	Yes	Yes
- due to combustion processes	Yes[2]	Yes	No	No	Yes
- due to direct containment heating	Yes	Yes	Yes	Yes	Yes
• Missiles or pressure loads					
-due to steam explosions	Yes[2]	Yes[2]	Yes[2]	Yes[3]	Yes[2]
• Meltthrough					
- due to direct contact between core debris and containment	No	Yes	Yes	No	No
• Vessel thrust force					
- due to blowdown at high pressure	Yes[2]	Yes[2]	No	No	No
Late Containment Failures					
• Overpressurization with high temperatures					
- due to noncondensible gases and steam	Yes	Yes	Yes	Yes	Yes
- due to combustion processes	Yes	Yes	No	No	No
• Meltthrough					
- due to basemat penetration by core debris	Yes	Yes	Yes	Yes	Yes
• Vessel structural support failure					
- due to core debris erosion	No	No	Yes	Yes	Yes

Notes:
    N/A = Not applicable.
    [1]Relatively low probability but potentially high consequences.
    [2]Low probability.
    [3]Possibility of steam explosion in downcomers of some Mark II designs.

(Based on information provided in Ref. A.3)

Some analysts prefer to use one generalized CET that covers most of the plant damage states, whereas other analysts use different CETs for the various plant damage states. The approach is left to the preference of the analyst. However, if one generalized CET is used, enough information ought to be given in Time Period 1 to adequately define the events in subsequent time periods. For example, whether the primary system is initially at high pressure (transient initiator) or low pressure (large-break LOCA initiator) can have a significant influence on subsequent in-vessel and ex-vessel phenomena.

In summary, many CETs have been developed over the last several years for a variety of reactor and containment designs (refer to Appendix B). The staff encourages the use of these CETs for performing an IPE if they can be shown to be appropriate. If it is necessary to develop CETs as part of the IPE, the following should be noted:

- Select event tree headings, such as those given in NUREG/CR-2300.
- Divide the accident into major time periods, such as those given above.
- Divide events related to both in-vessel and ex-vessel phenomena into:

    - High-pressure sequences, and
    - Low-pressure sequences.

Step 5 - Determination of Containment Challenges and Time of Failure

In this approach, containment challenges would be determined for each of the representative sequences identified for the various plant damage state bins (Step 2 above). In this context, "challenges" refers to the potential for elevated pressures and temperatures, missiles, direct contact of containment by core debris, containment bypass, and the like that could be caused by the core meltdown accident. The magnitude of these challenges when compared with the containment capacity (Step 3) will determine if containment failure will occur and, if it does, the time at which failure is reached. This information is therefore extremely important and is needed to quantify the CETs (Step 7).

During the last several years, there have been extensive evaluations of containment challenges during severe accidents for a variety of reactors (refer to Appendix B). The staff encourages the use of these existing calculations whenever they can be shown to be applicable. Again, if calculations are to be used from industry PRAs, the staff review should also be taken into account. Perhaps the most up-to-date and extensive assembly of information related to containment challenges is provided in the latest version of NUREG-1150 and particularly in the supporting contractor reports. Reports of particular relevance to containment challenges are NUREG/CR-4551, Volumes 1 through 7 (Refs. A.7 through A.13). Volume 2 of NUREG/CR-4551 (Ref. A.8) provides distributions of containment challenges that were developed by experts drawn from national laboratories, universities, and industry. This information is provided in the form of distributions; the analyst can extract the appropriate information (mean, median, etc.) for use in the plant examination. The analyst should also be aware of the associated uncertainty range (as reflected in the distributions), and these ranges can help form the basis for a sensitivity study (Step 8). It should be noted that licensees need not explicitly calculate plant-specific distributions for the IPE.

Two specific containment issues were identified in Generic Letter No. 88-20 (Ref. A.14): direct containment heating (DCH) and liner meltthrough. Both of these issues were considered in Volume 2 of NUREG/CR-4551 (Ref. A.8) and contribute to the uncertainty distributions. Thus, if an analyst chooses to use the information in Volume 2 of NUREG/CR-4551, related to containment pressure and temperature loads resulting from core meltdown accidents with the primary system at high pressure, these distributions include consideration of DCH. If the analyst wishes to perform sensitivity studies (Step 8), the use of the high end (95th percentile confidence level) of the distributions will provide an estimate of the vulnerability of the containment being examined to DCH phenomena. The above discussion also applies to the liner meltthrough concern for Mark I containments. If the information in NUREG/CR-4551 is used, the uncertainty ranges include consideration of this issue.

Step 6 - Determination of Source Term Magnitude

By comparing estimates of the magnitude of the containment challenges (Step 5) with the capability of the containment to withstand these challenges (Step 3), the timing and mode of containment failure or bypass can be determined. After this process is completed, descriptions of the various potential fission product release paths are obtained and therefore the timing, magnitude, and character-istics of accidental radionuclide releases can be determined.

During the last several years, there have been extensive evaluations of fission product release (source terms) during severe accidents for a variety of reactor designs (refer to Appendix B). The staff encourages the use of these existing calculations whenever they can be shown to be applicable. Consideration must be given to the types of sequences in the release category, however, and the timing on release characteristics for each before selecting release character-istics to represent the category.

The Reactor Safety Study (RSS) (Ref. A.15) was an early attempt to estimate severe accident source terms. However, the RSS methods contain simplifications that may tend to overestimate or underestimate the magnitude of some radionuclide species for some accident sequences. After the publication of the RSS, signif-icant research was undertaken to better define severe accident source terms. Updated source terms methods were developed for the NRC and published in BMI-2104 (Ref. A.16). A technical assessment of severe accident source term technology was published in NUREG-0956 (Ref. A.17). This assessment reviewed experimental and analytical results from severe accident research and recommended the Source Term Code Package (STCP) as a viable tool for source term evaluation, provided uncertainties were considered. An extensive series of STCP calculations (Ref. A.18) for various accident sequence and reactor designs formed the basis for the source term estimates in draft NUREG-1150 (Ref. A.5). The STCP calcula-tions in Reference A.18 can be used in an IPE if a source term can be identified that was calculated for similar accident progression characteristics and reactor design. Whatever calculational method is used, the analyst should be aware of the inherent uncertainties. The uncertainty ranges in draft NUREG-1150, for example, can be used to guide the sensitivity studies (Step 8).

Step 7 - Quantify CETs

Quantification of a CET involves allocating probabilities to each of the branches (or node points) in the tree. These probabilities are based on the analyst's understanding of the events under consideration. For example, the analyst would use (1) referenced plant information, (2) developed distributions, or (3) sensitivity studies for the pressure rise at vessel breach and for containment failure pressure. Then, by comparing the ranges for pressure rise and containment failure, a split fraction can be estimated for the branch point (e.g., 0.1 for failure and 0.9 for no failure). This process is repeated for each of the branches in the CET. The probability of each path through the CET (or endpoint) can then be determined by multiplying all the branch point probabilities but taking into account any dependencies from earlier questions in later questions. This later point is especially important for larger event trees. All the endpoint probabilities in the CET should sum to unity for each front-end bin.

Quantification of the CETs will produce estimates of the probabilities of various containment failure modes, bypass events, or no containment failure for the plant damage states identified (Step 2). By combining the frequencies of the plant damage states with the probabilities of the CET endpoints, the frequencies of containment failure or bypass can be determined. Each failure mode identified for each plant damage state has a unique fission product release characteristic (or source term). It is common practice to reduce the large number of possible source terms to a smaller number of representative release categories. In the RSS (Ref. A.15), nine PWR release categories and five BWR release categories were identified. However, as our understanding of source term phenomenology increased, the number of representative release categories used in some recent PRAs was increased to better represent the range of possible source terms. For example, in draft NUREG-1150 (Ref. A.5), many more release categories were identified than in the RSS.

The number of representative release categories selected is left to the discretion of the analyst; however, it is essential to select a sufficient number of representative release categories so that each of the individual source terms can be adequately represented by one of the representative release categories.

After a set of representative source terms has been established, each of the CET endpoints can be allocated to an appropriate representative source term. The frequency of any given representative source term can then be determined by summing all the CET endpoint frequencies for each of the plant damage states that are allocated to it. When this process is complete, the relative importance (magnitude of source term or frequency) of each containment challenge can be determined.

Step 8 - Sensitivity Studies

Steps 2 through 7 above are subject to various forms of uncertainty, and Appendix 1 to the Generic Letter also states that CET quantification should include consideration of uncertainties. There are various ways of propagating uncertainties through the back-end portion of a PRA. Draft NUREG-1150 presents the most extensive attempt to propagate uncertainty. However, the NRC staff does not believe that it is necessary to use a method as sophisticated as the

NUREG-1150 approach for the purpose of an IPE. A well-structured sensitivity study ought to be sufficient to determine what has the largest effect on the likelihood or time of containment failure and the magnitude of the source term without calculating the uncertainties explicitly.

In order to perform a sensitivity study, those parameters that are likely to have the largest effect must be identified. Appendix 1 to the Generic Letter provides a discussion of those parameters that have been found to have a significant effect on containment failure and source terms in past studies. These parameters are summarized in Table A.5. The parameters in Table A.5 represent a reasonably comprehensive list of parameters for use in a sensitivity study. However, it may be necessary to subtract or add to the list depending on the configuration of the containment under consideration.

The staff encourages the use of previous studies (refer to Appendix B) to help determine the feasible ranges for the various parameters in Table A.5 (e.g., the effect of the core debris being coolable or not coolable). The information provided in Volume 2 to NUREG/CR-4551 (Ref. A.8) represents the most up-to-date information on the uncertainty associated with the parameters in Table A.5 and can be used to help estimate the limits of the ranges for the purposes of the sensitivity study.

After the parameters are identified and the ranges established, the CETs can be requantified by varying the parameters over the various ranges to determine those parameters that have the largest effect on containment failure and source term magnitude. This process identifies those areas for which potential improvements might be considered or could indicate how robust conclusions about vulnerabilities are in the face of the uncertainties, provided that uncertainties are comparable to the established ranges.

Table A.5 Parameters for sensitivity study.

- Performance of containment heat removal systems during core meltdown accidents

- In-vessel phenomena (primary system at high pressure)

  - H_2 production and combustion in containment
  - Induced failure of the reactor coolant system pressure boundary
  - Core relocation characteristic
  - Mode of reactor vessel meltthrough

- In-vessel phenomena (primary system at low pressure)

  - H_2 production and combustion in containment
  - Core relocation characteristics
  - Fuel/coolant interactions
  - Mode of reactor vessel meltthrough

- Ex-vessel phenomena (primary system at high pressure)

  - Direct containment heating concerns
  - Potential for early containment failure due to pressure load
  - Long-term disposition of core debris (coolable or not coolable)

- Ex-vessel phenomena (primary system at low pressure)

  - Potential for early containment failure due to direct contact by core debris

  - Long-term core-concrete interactions:

    o Water availability
    o Coolable or not coolable

REFERENCES FOR APPENDIX A

A.1   J. W. Hickman, "PRA Procedures Guide: A Guide to the Performance of Probabilistic Risk Assessments for Nuclear Power Plants," American Nuclear Society and Institute of Electrical and Electronic Engineers, NUREG/CR-2300, Vols. 1 and 2, January 1983.

A.2   L. D. Bustard et al., "EQ Risk Scoping Study," Sandia National Laboratories, NUREG/CR-5313, SAND88-3330, January 1989.

A.3   Brookhaven National Laboratory, "Assessment of Severe Accident Prevention and Mitigation Features," NUREG/CR-4920, Vols. 1-5, BNL-NUREG-52070, July 1988.

A.4   Industry Degraded Core Rulemaking (IDCOR) Program, "Individual Plant Evaluation Methodology for LWRs," April 1987.

A.5. USNRC, "Severe Accident Risks: An Assessment for Five U.S. Nuclear Power Plants," NUREG-1150, Vols. 1 and 2, Second Draft for Peer Review, June 1989.

A.6   Letter from A. Thadani, NRC, to W. Rasin, NUMARC, "Staff Evaluation of IDCOR IPEMs," dated November 22, 1988.

A.7   E. D. Gorham-Bergeron et al., "Evaluation of Severe Accident Risks: Methodology for the Accident Progression, Source Term, Consequence, Risk Integration, and Uncertainty Analyses," Sandia National Laboratories, NUREG/CR-4551, Vol. 1, Draft Revision 1, SAND86-1309, to be published.*

A.8   F. T. Harper et al., "Evaluation of Severe Accident Risks: Quantification of Major Input Parameters," Sandia National Laboratories, NUREG/CR-4551, Vol. 2, Draft Revision 1, SAND86-1309, to be published.*

A.9   R. J. Breeding et al., "Evaluation of Severe Accident Risks: Surry Unit 1, Sandia National Laboratories, NUREG/CR-4551, Vol. 3, Draft Revision 1, SAND86-1309, to be published.*

A.10 A. C. Payne, Jr., et al., "Evaluation of Severe Accident Risks: Peach Bottom Unit 2," Sandia National Laboratories, NUREG/CR-4551, Vol. 4, Draft Revision 1, SAND86-1309, to be published.*

A.11 J. J. Gregory et al., "Evaluation of Severe Accident Risks: Sequoyah Unit 1," Sandia National Laboratories, NUREG/CR-4551, Vol. 5, Draft Revision 1, SAND86-1309, to be published.*

A.12 T. D. Brown et al., "Evaluation of Severe Accident Risks: Grand Gulf Unit 1," Sandia National Laboratories, NUREG/CR-4551, Vol. 6, Draft Revision 1, SAND86-1309, to be published.*

---

*Available in the NRC Public Document Room, 2020 L Street NW., Washington, DC.

A.13 C. K. Park et al., "Evaluation of Severe Accident Risks: Zion Unit 1," Brookhaven National Laboratory, NUREG/CR-4551, Vol. 7, Draft Revision 1, BNL-NUREG-52029, to be published.*

A.14 NRC letter to All Licensees Holding Operating Licenses and Construction Permits for Nuclear Power Reactor Facilities, "Individual Plant Examination for Severe Accident Vulnerabilities - 10 CFR §50.54(f)," Generic Letter No. 88-20, dated November 23, 1988.

A.15 USNRC, "Reactor Safety Study--An Assessment of Accident Risks in U.S. Commercial Nuclear Power Plants," WASH-1400 (NUREG/75-014), October 1975.

A.16 J. A. Gieseke et al., "Radionuclide Release Under Specific LWR Accident Conditions--PWR Large, Dry Containment Design (Surry Plant Recalculations)," Battelle Columbus Laboratories, BMI-2104, Vol. V, Draft, July 1984.

A.17 M. Silberberg et al., "Reassessment of the Technical Bases for Estimating Source Terms," USNRC Report NUREG-0956, July 1986.

A.18 R. S. Denning et al., "Report on Radionuclide Release Calculations for Selected Severe Accident Scenarios," Battelle Columbus Laboratories, NUREG/CR-4624, Vols. 1-5, BMI-2139, July 1986.

*Available in the NRC Public Document Room, 2120 L Street NW., Washington, DC.

APPENDIX B

PRA REFERENCES

Tables contained in this appendix list PRAs either performed or reviewed by the NRC. If a licensee chooses to use calculations that were performed in support of a PRA listed in this appendix, they should also take into account the findings of the NRC staff's review.

Table B.1 PRAs done by NRC.

Analyst	Plant	Type	PRA level	Report No.	Comment
SNL	ANO-1	B&W	2	NUREG/CR-2787	IREP
SNL	ANO-1	B&W	3	NUREG/CR-4713[I]	TAP A45
INEL	Browns Ferry 1	BWR4 MK1	1	NUREG/CR-2802[I]	IREP
SNL	Calvert Cliffs 1	CE	1	NUREG/CR-3511	IREP
SNL	Calvert Cliffs 2	CE	1	NUREG/CR-1659	RSSMAP
SNL	Cooper	BWR4 MK1	3	NUREG/CR-4767[I]	TAP A45
SAI, SNL	Crystal River 3	B&W	2	NUREG/CR-2515[I]	IREP
SNL	Grand Gulf 1	BWR6 MK3	1	NUREG/CR-1659[I]	RSSMAP
SNL	Grand Gulf 1	BWR6 MK3	3	NUREG/CR-4550, 4551, 4700, 4624[I]	NUREG-1150
SNL	LaSalle 2	BWR5 MK2	3	Unavailable	RMIEP (in progress)
NEU	Millstone 1	BWR3 MK1	1	NUREG/CR-3085[I]	IREP
SNL	Oconee 3	B&W	2	NUREG/CR-1659[I]	RSSMAP
	Peach Bottom 2	BWR4 MK1	3	WASH-1400[I]	RSS
SNL	Peach Bottom 2	BWR4 MK1	3	NUREG/CR-4550, 4551, 4700, 4624[I]	NUREG-1150
SNL	Point Beach 1	W2	3	NUREG/CR-4458	TAP A-45
SNL	Quad Cities 1	BWR3 MK1	3	NUREG/CR-4448[I]	TAP A-45
SNL	Sequoyah 1	W4 IC	1	NUREG/CR-1659[I]	RSSMAP
SNL	Sequoyah 1	W4 IC	3	NUREG/CR-4550, 4551, 4700, 4624[I]	NUREG-1150
SNL	St. Lucie 1	CE	3	NUREG/CR-4710	TAP A-45
SNL	Surry 1	W3 SA	3	NUREG/CR-4550, 4551, 4700, 4624[I]	NUREG-1150
AEC	Surry 1	W3 SA	3	WASH-1400	RSS
SNL	Turkey Point 1	W3	3	NUREG/CR-4762	TAP A-45
INEL/BNL	Zion	W4	3	NUREG/CR-4550, 4551, 4700, 4624[I]	NUREG-1150

I = Another PRA on the same plant sponsored by the industry.

Note: IREP   - Integrated Reliability Evaluation Program
       TAP   - Task Action Plan
       RSSMAP - Reactor Safety Study Methodology Application Program
       RMIEP   - Risk Methodology Integration and Evaluation Program
       RSS   - Reactor Safety Study

Table B.2 Industry PRAs reviewed or under review by NRC staff.

Analyst	Plant	Type	PRA level
CP	Big Rock Point[N]	BWR1	3
PLG	Browns Ferry 1[N]	BWR4 MK1	3
EII	Brunswick 1	BWR4 MK1	1
EII	Brunswick 2	BWR4 MK1	1
SAIC	Crystal River 3[N]	B&W	1
PLG	Diablo Canyon	W4	1
GE	GESSAR II	BWR	3
NEU	Haddam Neck	W	1
PLG, W. Fauske	Indian Point 2	W4	3
PLG, W. Fauske	Indian Point 3	W4	3
NUS, GE	Limerick 1	BWR4 MK2	3
NUS, GE	Limerick 2	BWR4 MK2	3
NEU	Millstone 1[N]	BWR	1
NEU	Millstone 3	W4	3
EPRI, Duke Power	Oconee 3[N]	B&W	3
PLG	Seabrook	W4	3
SAI	Shoreham	BWR4 MK2	3
B&W, PLG	TMI 1	B&W	3
EII, YAEC	Yankee Rowe	W4	3
PLG	Zion[N]	W4	3

N = Plant also PRAd by the NRC.

Table B.3 Reports by NRC on industry PRAs.

Analyst	Plant	Type	PRA level	Report No.	Comment
INEL	Big Rock Point	BWR	3	EGG-EA-5765	
INEL	Brunswick	BWR4 MK1	1	Unavailable	In progress
ANL	Crystal River 3	B&W	1	NUREG/CR-3245	
BNL	Diablo Canyon	W	1	Unavailable	In progress
SAIC	Haddam Neck	W4		NUREG-1185	
SNL	Indian Point	W4	3	NUREG/CR-2934	
BNL	Limerick	BWR4 MK2	3	NUREG/CR-3028, 3493, 1068	
BNL	Midland	B&W	3	BNL Tech. Report A-3777	
SAIC	Millstone 1	BWR2 MK1		NUREG-1184	
NRC	Millstone 3	W4	3	NUREG-1152	
LLNL	Millstone 3	W4	3	NUREG/CR-4142	Level 1 review
BNL	Millstone 3	W4	3	NUREG/CR-4143	Levels 2 & 3 review
BNL	Oconee 3	B&W	3	NUREG/CR-4374 Vols. 1/2	Level 1 review
			3	NUREG/CR-4374 Vol. 3	Levels 2 & 3 review
LLNL	Seabrook	W4	3	NUREG/CR-4552	
BNL	Shoreham	BWR4 MK2	3	NUREG/CR-4050	
BNL	Yankee Rowe	W4	3	NUREG/CR-4589	
SNL	Zion	W4	3	NUREG/CR-3300	

APPENDIX C

NRC RESPONSE TO COMMENTS AND QUESTIONS

INTRODUCTION

On December 29, 1988, a Federal Register notice (53 FR 52881) announced that an Individual Plant Examination (IPE) workshop would be held in Fort Worth, Texas, on February 28 and March 1-2, 1989, to discuss the IPE objectives and solicited questions and points for clarification on the draft of this document. A February 8, 1989 Federal Register notice (53 FR 6184) provided the public with a preliminary workshop agenda and announced the availability of the draft.

This appendix paraphrases, summarizes, and categorizes into subject areas questions and comments that stem from the IPE workshop. These questions and comments were either raised at the workshop or were received by the staff (11 parties submitted written comments) soon afterwards. The NRC staff response is also provided. Table C.1 contains a listing of the subject areas discussed in this appendix. The workshop transcript and a copy of the comments that were received in writing are available in the NRC Public Document Room.

1.   IPE PROGRAM INTEGRATION AND RELATIONSHIP OF CPI TO IPE

1.1   With regard to the relationship between the Containment Performance Improvement (CPI) program and the Individual Plant Examination (IPE), it appears that the staff is telling the utilities to hold off on plant-specific IPE containment fixes until the generic CPI effort is completed. But in the case of the Mark I containment, it appears that a different meaning is intended. There could be several mandated generic fixes coming up for the Mark I before the IPE is completed, although the IPE may determine that the Mark I fixes are not required.

Response - The CPI and IPE programs are two major elements of an integrated plan (SECY-88-147) for severe accident closure. The CPI effort is based on the conclusion that there are known generic severe accident challenges to each containment type (e.g., overpressurization of the containment from sequences such as long-term loss of decay heat removal and challenges to the containment boundary from molten core debris) that should be assessed to determine whether additional regulatory guidance or requirements are warranted. In contrast, the purpose of the IPE program is to identify vulnerabilities that are unique to plants (e.g., system capacities and dimensions, valve alignments, and procedures) and that would not be found without a systematic examination of each plant. The staff has scheduled its efforts on the CPI program to provide its findings well in advance of the expected completion of the IPE effort by most licensees. Therefore, most will have the opportunity to factor the results of that program, including required implementation, if any, into their IPEs. The CPI program is described in SECY-88-147 (Ref. C.1). The staff briefed the Commission on recommendations for the Mark I containment improvement program on January 26, 1989, and the staff plans to provide additional guidance on CPI for other containment types by the end of January 1990.

Table C.1 Categorization of questions and comments.

1. IPE Program Integration and Relationship of CPI to IPE

2. Back-End Analyses

3. IDCOR IPE Methodology (IPEM)

4. Screening Criteria and Sequence Breakdown

5. External Events and Internal Flooding

6. Modes of Operation

7. Confirmation of the "As-Built As-Operated" Plant

8. Resources Needed to Perform the IPE

9. Treatment of Human Factors

10. Data, Uncertainty, and Treatment of Common-Cause Failure

11. IPE Documentation and Submittal Format

12. Vulnerabilities and Treatment

13. Consideration of Unresolved Safety Issues and Generic Safety Issues

14. NRC Staff Review and Review Guidance

15. Independent Review of the IPE

16. Equipment Survivability

17. Staff Response to IPE Submittals

18. Emergency Operating Procedures (EOPs)

19. Accident Management

20. Operator Training

21. Accident Strategies

22. Application of 10 CFR 50.59 Criteria to Severe Accidents

23. Integrated Safety Assessment

24. Regional Inspections

25. General Comments and Questions

1.2   The staff has talked about modifications in three elements: accident
         management, generic containment improvements, and IPE. They are all
         resource intensive. Industry is concerned that not enough thought
         has gone into coordinating and integrating these three programs.

Response - The staff has produced a document called the Severe Accident
Integration Plan and will be working closely with NUMARC on implementation.
Senior NRC managers are involved in this plan and are aware of the need for
proper integration.

1.3   We do not believe the CPIs for the Mark I are generic. They would be
         best handled as part of the IPE.

Response - The word generic applies to the word "vulnerability." It should be
recognized that fixes or modifications are proposed or recommended in a plant-
specific manner (i.e., taking into consideration plant-unique features), but
the vulnerabilities that the staff is looking at in the CPI program are
generic.

1.4   The IPE may require plant changes. Some may be an asset, but an
         integrated change may actually require that a previous commitment to
         the NRC (or installed system) may have to be deleted at the same time
         as a change is made in order to make use of the change. Will the
         staff have enough resources to process changes, or will the NRC be
         backlogged in having to approve deletions of previous improvements?

Response - It is difficult to project the impact of this potential "backlogged"
issue. We encourage the industry to come forward, and the staff will apply the
resources needed to approve the changes. The situation could apply either on
a plant-specific basis or on a generic basis. The staff would prefer to deal
with the changes in some generic fashion to maximize the efficiency of resources.
In addition, licensees may want to consider the benefits of the ISA program in
prioritizing and resolving issues. The staff would assign high review
priority, consistent with our procedure, to issues of high safety significance.

1.5   We suggest that the staff revise their approach to dealing with the
         back-end issue and allow the IPEs to proceed to full completion
         before requesting generic fixes to perceived containment problems.

Response - The staff intends to complete and identify containment design
recommendations (which stem from the Containment Performance Improvement
program) to the Commission by the end of 1989. The staff expects that
utilities affected by the execution of those recommendations will have ample
opportunity to factor them into their IPE evaluations.

2. BACK-END ANALYSES

2.1   There appears to be confusion over the focus of the back-end analyses. Should the focus be on release timing and containment failure modes or on source term or release magnitude?

Response - The primary focus of the back-end analyses should be on containment failure mode and release timing. However, releases are associated with containment failure so that, once the hierarchy of containment failure modes and timing have been developed, existing information can be used to determine the release magnitude for the various release categories. The associated releases can be derived from existing information. Detailed calculations may not be necessary.

One of the basic reasons for focusing on containment failure mode and timing is that it can immediately make obvious the type of response that can either mitigate or reduce containment failure probability. The text has been clarified to reflect this view. (See Section 2.2.2.7.)

2.2   What is meant by "template?"

Response - Template means existing PRA information or models. For example, in some cases front-end systems on a plant under study may be sufficiently similar to another (referenced) plant that the referenced plant's fault trees (or "templates") can be used as a starting point. For the back-end, the containment design, systems transient response, and failure modes of a plant under study may be sufficiently similar to that of a referenced plant that the reference plant ("template") analysis can be used instead of extensive code calculations.

2.3   How does the IPE analyst quantify direct containment heating and liner meltthrough. Does the staff expect quantification?

Response - The staff does not expect quantification of direct containment heating and liner meltthrough sequences because of the large uncertainties associated with these two issues. The analyst, however, should be aware of the range of possibilities, the uncertainties, and should allow for potential response actions under the accident management program. Any potential changes to the containment systems that stem from these two issues will be determined in the CPI program. Further insights on the back-end analysis can be obtained from Appendix A.

2.4   Vapor explosions were listed on one of the staff's workshop viewgraphs although consensus in the industry indicated that alpha-failure modes were considered to be of very low probability. Does that mean that there are other types of conditions that the analysts need to consider?

Response - The viewgraph in question listed containment failure modes considered in NUREG-1150 (Ref. C.2). The Generic Letter stated that vapor explosions themselves are not unlikely, although NUREG-1150 found the alpha-failure mode is not likely. Vapor explosion and alpha failure should not be used interchangeably.

2.5   Using different codes for the same accident sequence can lead to
        different answers. A source term calculated with one code can be
        well above the cutoff for the BWR-3 or PWR-4 that needs to be
        reported, while with another code it is well below. It appears that
        one could take a reasonable position and end up with not having to
        report any source term.

Response - Different code inputs could also provide different source terms.
Letters referenced in this document provide the NRC position on phenomenological
issues. Following the staff's recommendation on those issues could go a long
way in eliminating code output differences. Source terms should be estimated
for all sequences reported in response to the core damage frequency screening
criteria.

2.6   Why report source terms at all if the staff knows from the descrip-
        tion of the accident sequence that one source term, for example,
        containment bypass, is more severe than another? A documented source
        term that can be very questionable is not needed to make the point.

Response - The emphasis on the back-end should be on containment failure mode
and timing rather than on source term analysis. Accident management, however,
cannot ignore the release and transport of both radioactive material inside
containment and that released to the environment.

2.7   It is pretty difficult to learn from different phenomena without a
        perspective on what the controlling physics are in the problem. It
        would be appropriate to look at available experiments and see how
        they relate to specific conditions or systems that one might have at
        a plant. It is difficult to have utilities look at their contain-
        ments when the staff has not told them what they should look for in
        terms of the subcompartments of the containment and the configuration.
        Experiments have been done. The staff ought to consider allowing the
        utilities to do something constructive that would contribute to the
        whole information base and have themselves learn whether or not the
        physical processes we've been discussing in broad generalities are
        real or just a figment of some people's imagination.

Response - The staff is always willing to consider judgments that were based on
sound experimental evidence.

2.8   Is it possible to use the existing MAAP analysis coupled with the
        existing NUREG-1150 information, coupled with other information, and
        develop a simplified methodology that accounts for all pertinent
        phenomena but would not require a massive code analysis?

Response - The ability of codes to perform a realistic accident progression
calculation depends upon the assumptions and judgment of the analysis team. It
is therefore not possible to predetermine the credibility of a simplified
methodology over a spectrum of accident conditions without knowledge of the
simplifying assumptions and physical conditions.

2.9   For plants that may not have the ability or the resources to do plant-specific source term analysis using available codes, are there sources publicly available, in addition to those referenced in [draft] NUREG-1335, that would lead these plants to have results more in line with those found for a comprehensive analysis?

Response - Appendix A and Appendix B now contain insights and sources of this information.

2.10   Not until all the containment systems are put into the plant model will the analyst be able to trace the shared dependencies between the containment systems and plant systems. Plant-specific containment structural analysis can be quite significant if one is trying to address leak-before-break issues. Leak-before-break issues cannot be addressed with generic calculations for another containment because they depend on large strains, large deformations, and interferences that result from those deformations. Otherwise generic calculations might be quite appropriate.

Response - Because of the many types of containment structures, it is important that the analyst be confident that any referenced structural calculations used in the analysis reflect the design under examination. This is extremely important for containment studies that include credit for leak-before-break.

2.11   Accident management has to consider potential core damage progression. What codes are acceptable for this type of analysis? EPRI is making a real effort to do something to MAAP by incorporating work that was done in the DOE ARSAP program and putting BWR SAR into it. I think this will be a real improvement, but does the staff find it acceptable?

Response - The code analysis may or may not be acceptable. The specific code to be used is not so important as the consideration of the full range of phenomenological outcomes. It is important for utilities to understand where the uncertainties lie and to know how those uncertainties can affect what measures can or should be taken following an accident. Volume 2 of NUREG/CR-4551 (Ref. C.3) provides the most up-to-date and comprehensive discussion of these phenomena.

2.12   The staff should state in NUREG-1335 that it will not take a position on the acceptability of, or the relative merits of, the various accident analysis codes or human error rate methodologies used in the IPE.

Response - The purpose of this document is to provide IPE submittal guidance for utilities and not to publish positions on various issues. Staff positions can be found in the referenced documentation and previous PRA reviews.

2.13   The Generic Letter specifically states that the first node of the containment event tree should be a question related to containment isolation. This seems unduly prescriptive, especially because of the concern associated with accounting for dependencies between active systems appearing in the back-end event trees.

Response - The staff agrees that requiring the first nodal decision point of the containment event tree as a question related to containment isolation seems unduly prescriptive. Past PRAs have treated containment isolation either in the front-end systems analysis or the containment analysis and need not be treated differently (or specifically revised for already completed PRAs) as part of the IPE. In either case, the analyses should address those areas and contributors identified in the Generic Letter. Section 2.2.2.4 has been modified to reflect this view.

2.14   Blowdown forces/vessel thrust force are no longer considered an issue and have not been included in draft NUREG-1150. Failure modes and mechanisms not included in draft NUREG-1150 should not be included in the guidance document and should not be required to be part of the IPE process.

Response - Reference should be made to the 1989 draft of NUREG-1150 (Ref. C.2). Blowdown forces/vessel thrust forces were considered.

2.15   Section 2.2.2.7 in NUREG-1335 should be changed to reflect our understanding of the staff intent that the utility is responsible for identifying important sequences and vulnerabilities. Thus, the utility will decide which sequences require estimation of the magnitude of the radionuclide release.

Response - All sequences found to be important should have an estimation of the magnitude of the radionuclide release. The screening criteria help to identify these important sequences and a source term should be estimated for those sequences with a core damage frequency of 10[-6] per year or greater for functional sequences and 10[-7] per year or greater for systemic sequences. Other sequences may also be identified by the utility as important. The need for the estimation of the radionuclide release for these sequences should be evaluated on a case-by-case basis by the utility.

2.16   Is it not reasonable to assume that the NRC staff means that large [release] is greater than or equal to BWR-3 or PWR-4...?

Response - At the time of this writing, the staff has not specifically defined "large release." Options are being considered for defining a large release and plant performance objectives as part of the safety goal objectives.

2.17   Many of the insights typically gained in performing a plant-specific containment analysis will not be acquired by the utility team if they choose to perform little or no plant-specific analysis and elect instead to use prior analysis results.

Response - By electing to choose a reference plant analysis where similarity exists, the staff expects that many severe accident insights can be gained without requiring that each plant perform a containment phenomenological analysis. Furthermore, plant-specific containment system analyses are required as part of the IPE.

2.18   In the consequence analysis, there is a wide variability associated with the acceptability of an answer or having that answer accepted by a large diverse audience. In addition, taking the most adverse time window for weather and subsequent consequence calculation will result in an answer that will be unrealistic for most of the year. In what sense does the NRC staff consider consequences in determining the priorities and urgency of accident management strategy?

Response - First, codes like MACCS that do the consequence calculation choose randomly from a year's weather pattern by using Monte Carlo techniques. This represents the actual meteorological distribution of the plant rather than the worst case. Second, there are a number of things aimed at the prevention of core damage that one would want to do without getting into the details of the consequence modeling.

The real focus should be on those areas that can be controlled (via a combination of strategies that could be developed into procedures) and can reduce the likelihood of a large release. Carrying the IPE into a Level 3 PRA category in order to obtain some bottom line risk estimate would not significantly contribute to that reduction.

3.   IDCOR IPE METHODOLOGY (IPEM)

3.1 Utilities should be allowed to apply the IPEM without staff enhancements.

Response - Utilities should not apply the IPEM without the staff enhancements. The enhancements required by the staff to the "front-end" IPEMs play an important role in identifying vulnerabilities and are necessary to accomplish the stated objectives in the Generic Letter. These enhancements are intended to improve the IPEM and not eliminate it as an option for performing the IPEs.

3.2   The Generic Letter, while not changing the stated purpose of the evaluations, imposes major changes to the "front-end" methods established in the IPEM and appears to find the containment and source term methods unacceptable.

Response - The staff has repeatedly stated the belief that the IPEM (especially for BWR plants) without enhancements may only be capable of identifying vulnerabilities previously known from PRA experiences. Even though this in itself can be considered a desirable achievement, the staff believes that it is essential to have a methodology that would be capable of identifying vulnerabilities that may be unique to the plant under study.

The IDCOR IPEM containment and source term method was found unacceptable because it does not provide the necessary perspective to the utility to understand the progression of severe accidents, the roles and margin of available systems, and how accident management strategies could alter the course of the accident. In addition, the IPEM does not account for uncertainties and precludes several phenomena and alternative outcomes that have been recognized as plausible by the reactor safety community. Appendix 1 to the IPE Generic Letter provides guidance for evaluating containment performance. Further guidance is provided in Appendix A.

3.3   NUMARC assessed the Level 1 enhancements to the IPEM, as recommended by the NRC staff, to be purely minor in terms of resources and the overall process. Does the staff agree with this assessment?

Response - The staff enhancements would understandably increase the scope of the IPEM analysis. The increased scope will improve the ability of the method to estimate more reasonable and realistic plant vulnerabilities. The staff does not believe, however, that the enhancements will change the IPEM into a "full scope" Level 1 PRA. In PRAs, an in-depth method of generating all contributors is used, whereas in the IPEM, fault tree and event tree templates are used. These templates generate only numerical estimates for accident sequences (i.e., no component level cutsets are generated). The staff enhancements are therefore needed to expand or generate further the cutsets of (only) the outlier sequences in order to reveal the fundamental causes of the vulnerabilities. This should not significantly increase the resources needed to perform the IPEM.

3.4   Would the staff characterize the enhancement to the IPEM as being applicable to IPEM or PRAs in general?

Response - The enhancements that the staff proposed for the IPEM should also be concerns that need to be addressed by PRAs in general. From previous experience, however, most IPEM shortcomings identified in the staff's SER were found not to be a problem in PRAs.

3.5   If everyone addressed the enhancements with the PRA or IPE, will the staff be satisfied?

Response - A staff finding that the IPE is acceptable will be based on the staff's review and the extent to which the IPE met the Generic Letter 88-20 objectives.

3.6   Could the staff build on the list of support systems beyond those included in the IPEM?

Response - The staff's concern involves the judgment used in the IPEM to single out certain support systems as most important while leaving out others. All support systems should be considered.

The support systems are used in a separate part of the IPEM analysis, and the staff does not know how they are going to be processed through the quantification process. Some support systems are included in IPEM Appendix D, although the staff does not have any means of checking the validity of the IPEM Appendix D questions.

3.7   The IDCOR IPEM Safety Evaluation Reports and Appendix 1 to Generic Letter 88-20 strongly imply that nearly a full-scale Level 2 PRA (including full quantification with a containment structural analysis and addressing the spectrum of physical phenomena that can evolve in containment as a consequence of core melt) is necessary to satisfy the NRC staff.

Response - The staff believes that, in many cases, reference plant analyses can be used to achieve the insights that would otherwise result from performing a full-scale Level 2 PRA analysis. There are two fundamental guidelines in developing the "Appendix 1" approach for the containment performance part of the IPE that will help achieve this goal. First, the methodology encourages utilities to understand severe accident progressions, phenomenologies, and responses of systems (primarily containment) to these accidents. (The IDCOR IPEM did not do this to a satisfactory level.) Second, the methodology can be crafted in such a way that it need not rely on large computer codes and detailed "outside" analysis to reach that level.

3.8   Can utilities use IPEM Appendix D insights in performing their IPE?

Response - It is the utilities' option to use Appendix D insights; however, they should be aware that they must go further and develop fault trees in order to identify plant-specific vulnerabilities.

3.9   Is there a way to update the NRC staff and IDCOR issue resolution papers, given the status of research between now and then? Could the staff provide more specific guidance on three or four specific issues related to (at least) the BWRs?

Response - The conclusions of the NRC-IDCOR issue papers remain as valid today as when they were first written. In most cases, the conclusions are either that certain aspects of the issue are resolved or that a wide range of outcomes should be considered because of a paucity of experimental data or differing interpretations of existing data. If the papers were to be rewritten today, the review of the status (issue definition and staff assessment sections) would be updated to reflect the existence of additional research. However, the need for consideration of a wide range of outcomes for many phenomenological calculations is unchanged.

4.   SCREENING CRITERIA AND SEQUENCE BREAKDOWN

4.1   It is understandable why the staff requested that functional sequences be defined, i.e., so that sequences cannot be broken down to whatever low frequency one chooses by going down through the component and subcomponent level. However, one could define functions as simply shutdown and decay heat removal. Does the staff wish to have sequences accumulated to that high a level so that, for example, only two functional sequences be recorded, or will the staff provide guidance on specific safety functions for accumulation of the accident sequences?

Response - The functions are dependent on the type of plant and to some extent on the analyst's choice. The staff, however, does not believe the sequences should be limited to two or three. Some functional examples include for PWRs: reactor trip, RCS inventory control, decay heat removal, containment cooling, scrubbing of radioactive material, pressure control, and recirculation; for BWRs: reactivity control, high-pressure coolant makeup, low-pressure coolant makeup, containment heat removal, and depressurization.

4.2   We have no problem providing a description of the impact of each
           support system on each front-line system. In draft NUREG-1335,
           however, the staff requested the description of each support state
           and its effect on each front-line system. There are so many support
           states that the staff won't want them and we won't be able to do it.
           We have no problem providing each support system. We think that is
           more reasonable and more appropriate.

Response - Section 2.1.3 has been modified to request only those states (or
bins) found to be important.

4.3   The screening criteria found in Generic Letter 88-20 refer to an
           expected value. Is the expected value a mean or median?

Response - The mean value is the value that should be used.

4.4   One of the screening criteria in Appendix 2 says that the utilities
           should report "any functional sequence that contributes 1E-6 or more
           per reactor year to core damage," while another criterion states that
           "any functional sequence that has a core damage frequency greater
           than or equal to 1E-6 per reactor year and leads to containment
           failure which can result in a radioactive release magnitude greater
           than..." Are those mutually exclusive?

Response - They are not mutually exclusive. All functional sequences that
meet the $10^{[-6]}$/year or greater core damage criterion should be reported.

4.5   If a sequence has a core damage frequency and containment failure
           proba-bility less than the screening criteria, then the source term
           for that sequence need not be calculated. During the event tree
           process, can a sequence be terminated at a certain point if it falls
           below the screening criteria contained in the Generic Letter?

Response - If the front-end or back-end does not trip any of the screening
criteria, then a containment analysis for that sequence does not need to be
reported. In many cases, utilities will find that it is not prudent to
truncate simply because they fall below the screening criteria, however.

4.6   The scenarios that are most important for core melt may not
           necessarily be the most important for offsite consequences. It is
           therefore suggested that the highest frequency scenarios constituting
           a significant fraction of total core damage frequency be reported,
           instead of imposing an arbitrary cutoff.

Response - Systemic screening criteria, introduced in Section 2.1.6, has been
based on this principle. If functional sequences have been used, however, then
Appendix 2 (Generic Letter 88-20) screening criteria are applicable.

4.7   Can Generic Letter Appendix 2 Criterion 3 be interpreted to mean
           that, whenever a node is passed through in the containment event tree
           that drops the sequence below $10^{[-6]}$/year, the analyst no longer has
           to consider the source term, either timing or release magnitude, for
           that sequence?

Response - The staff does not mean to imply that important sequences that challenge the containment or plant be truncated. The sequences and screening criteria included in Appendix 2 are for reporting purposes only. A source term should be reported for all functional sequences with core damage frequency at or above 10[-6]/year (10[-7] for systemic sequences).

4.8   There are several rules that apply to initiating events (e.g., ATWS and station blackout). Utilities are diligently addressing or satisfying them. Could utilities exclude these events, or reduce the scope of the analysis, if these rules are satisfied or, for example, if the initiating events are below certain criteria or numbers?

Response - The staff has learned from experience that in spite of rules and regulations there may be unique situations that could make an otherwise resolved issue significant. The fact that some rule or regulation has been met does not justify excluding the issue from further consideration in the IPE.

4.9   The draft NUREG-1335 seems to make little distinction between functional and systemic event trees and, therefore, functional and systemic sequences. We suggest that no distinction be made between functional or systemic sequences.

Response - The staff will accept systemic sequences as well as functional sequences as originally requested. There are differences in the screening criteria, however, which are now clarified in Section 2.1.6. The intent of having systemic screening criteria is to have reported sequences that would otherwise have been reported under the Generic Letter screening criteria, had those sequences been consolidated into functional form.

4.10   What is the definition of core damage for IPE? Is it core melt or fuel clad damage, or is it the 10 CFR 50.46(b) criteria?

Response - For purposes of the IPE, each of the above constitutes core damage or onset of core damage.

4.11   Should the back-end analysis include non-core-melt sequences? Also, should non-core release (e.g., waste gas tank, spent fuel) be considered?

Response - The IPE analyst should consider the impact of the back-end analysis on the front-end, e.g., sequences where containment failure can lead to core damage. For those sequences that do not involve core damage explicitly, the analyst should be confident in the model and aware of the uncertainties before concluding that a sequence is not important. For reporting purposes, non-core-damage sequences can be screened out. The staff does not expect non-core-damage releases to be included as part of the IPE.

4.12   The guidance document should state that any methodology that accurately accounts for support system dependencies is acceptable.

Response - This statement is implied and need not be included explicitly.

5. EXTERNAL EVENTS AND INTERNAL FLOODING

5.1   The staff requested that internal flooding be addressed at this time as part of the internal events. A case could be made, however, for the opportunity to take advantage of the resources that are needed to do both fires and flood at the same time. How favorably does the staff look at a suggestion that one defer the specific application of the flood analysis until the fire analysis is performed?

Response - The IDCOR IPEM includes internal flooding as an internally initiated event. Including flooding as part of the internal events should not significantly impact the resources necessary to perform the IPE. Recognizing the importance of internal flooding and the fact that the methodology is available while other external events are still under consideration, the staff requested that internal flooding be considered now as part of the internal-event analyses. The staff is willing to consider on a case-by-case basis, however, situations where the IPE has significantly progressed to the extent that including the internal flood analysis now as part of the internal events would be inefficient and, in effect, place an unnecessary burden on the resources needed to complete the IPE. Consideration will be given to these situations upon review of licensee submittal plans. Preliminary analyses (or references) that indicate that internally initiated flooding events are not significant contributors to core damage at the site should also be provided with the submittal plans.

5.2   The vulnerabilities and insights that have been derived from past external-event analyses are as striking as those derived from the internal events. Utilities have taken those vulnerabilities as seriously as those derived from the internal-event analyses. The question focuses on the basis for the staff's decision not to proceed with external events. Does the staff perceive serious shortcomings with the methodology that has been applied in the past? Is there insufficient staff experience with external-event analyses and reviews of PRAs that have included external events? Are there insufficient industry bases for performing these analyses and may that be the reason for deferment?

Response - The reasons for deferral of the external events are as follows:

1. The staff must still decide which external events need to be considered.
2. There are many methods, and the staff is still considering the possibility of developing more simplified methods.
3. There are a number of ongoing programs at NRC and industry that need to be coordinated, e.g., seismic programs.

The delay is only until the staff formulates firmer plans on the external events.

5.3   Have the external events been delayed because the staff has identified shortcomings in past methods? In the back-end of the Level 2 PRA, there are large phenomenological uncertainties, including which phenomena to consider in the IPE. The Level 2 analysis has proceeded whereas the external events have not. I really do not see a fundamental difference between the situations.

Response - The delay in treatment of external events in the IPE is not because the staff has identified deficiencies in the external-event methodologies.

NUREG/CR-2300 (Ref. C.4) has a two-page listing of all external events of which not all are significant. By identifying those that are, the staff hopes to provide a reduction in the required resources of both the staff and the utilities.

6. MODES OF OPERATION

6.1   Does resolution of USI A-45 include shutdown events, e.g., Modes 4 through 6, or just from power situations?

Response - For purposes of the IPE, only power operation and hot standby need to be considered. The document was revised to reflect this view.

6.2   For events initiating in Modes 1, 2, or 3, how long after the initiating events do decay heat removal system mission times need to be included?

Response - Because of limitations in modeling scenarios that extend over long periods of time, the nominal assumption of 24 hours is sufficient for the IPE.

6.3   While full-power scenarios are unquestionably more demanding, many low power and even cold shutdown scenarios can lead to rapid overpressurizations, reactivity excursions, unusual plant line-ups, etc.

Response - This issue has been previously raised as a generic issue. Issues raised as generic issues need not be evaluated as part of the IPE program. Each utility has the freedom to resolve such an issue within the IPE framework, although resolution of generic issues is not required as part of the IPE program.

7. CONFIRMATION OF THE "AS-BUILT AS-OPERATED" PLANT

7.1   What is meant by verifying that the analysis reflects the plant design and operation? Will we have to go back and verify all our design documentation in terms of its representing the plant? For example, the information necessary to convince yourself that a plant is "single failure proof" for certain systems is different from that for a PRA where the failure mode of particular equipment is needed. You can analyze the plant and convince yourself that it meets single failure requirements, but you may not be able to determine the failure modes of certain equipment with the available information.

Response - If certain susceptibilities are identified because the plant does not meet NRC requirements, be it single failure or whatever else, then the staff will require that the deficiency be corrected. The staff is not saying, however: "Do the study to show us how the plant meets existing regulations." The intent of the term "plant as is" is to be sure, for example, that a PRA performed 5 years ago reflects any modifications and design updates if it is to be submitted as part of the IPE, or that information from the FSAR, for instance, represents the plant. The intent of the term, however, is not to have utilities do design verifications.

7.2   It is important that the staff provide guidance in NUREG-1335 with respect to how one goes about confirming that the analyzed plant is the as-built as-operated plant. Clarification is needed on what is an acceptable process and how it is to be documented. Is a walk-through alone sufficient?

Response - The staff recommends a walkthrough in which specific features are verified. The walkthrough should be structured and consist of team members such as field engineers and PRA analysts familiar with the plant and plant systems.

7.3   It would be helpful if the scope and team requirements for the IPE plant walkthrough were to be provided. For our plants there have been a number of walkthroughs previously for other NRC requirements. The external events will probably require another type of investigation, another walkthrough. Anything that could be provided to better delineate what would be expected here would be helpful.

Response - A walkthrough does not mean a complete detailed inspection of plant system configuration or operational aspects, but rather ensures that the analysis team reflects properly the design and operational aspects of the plant. Walkthroughs have to be carefully planned and scheduled to maximize their impact on the analysis. It should be realized that walkthroughs are not a single effort, but rather an iterative process, the extent of which is driven by the analyst's needs. The following list identifies the types of walkthroughs and personnel that might be considered:

- Initial walkthrough for plant familiarization.
- Special walkthrough for verification of logic trees or investigation of dependencies or aspects of system interactions.
- Each should have a team of plant personnel, PRA analysts, and any extra expertise compatible with the objective of the walkthrough (e.g., human factors, failure data analysis, electrical or instrument and control).
- External-event expertise may be required at a later stage (e.g., seismic, structural, fire, flood).
- Each walkthrough should be preplanned, and each member should be given an assignment to document results.

8. RESOURCES NEEDED TO PERFORM THE IPE

8.1   A clarification of the staff's person-hour estimates might be appropriate. It is probably true that an experienced PRA analysis team can provide the staff with the kind of analysis and documentation that has been requested in 8,000 to 9,000 person-hours. But the 8,100 person-hours is probably inconsistent with what the guidance is requesting. The guidance is asking for a substantial contribution by the utility team for the benefit of the utility. When you account for the fact that those are inexperienced PRA analysts, that their efficiency of contribution in systems analysis and data analysis, at least initially, is less than the experienced analysts, the staff's conclusion ought to be that the total person-hours spent probably will exceed the guidance that you have stated.

Response - The staff noted a wide range of estimates for the IPE effort, from 2-3 person-years per plant (IDCOR-estimated level presented to the Advisory Committee on Reactor Safeguards) to 8-10 person-years (Northern States Power Company estimate presented at the IPE workshop, Fort Worth, Texas). The major source of uncertainty stems from the background and experience of the IPE team performing the analysis. Based on the in-house NUREG-1150 effort and outside PRA practitioners' estimates, we believe an experienced team should be able to perform a plant-specific IPE within 8,100 person-hours.

The staff recognizes the fact that some utilities may have minimal PRA experience. For those utilities, the IPE effort could exceed twice the 8,100 person-hour estimate.

8.2   NUREG/CR-2300 (Ref. C.4) estimates manpower for a Level 1 PRA (excluding external events) to range from 11,000 to 20,000 person-hours. This did not include the back-end containment evaluation. In the IDCOR IPEM SER, the staff noted that the IDCOR IPEM (with staff enhancements) is estimated by the staff to require a level of effort commitment equivalent to a Level 1 PRA. Therefore, a direct correlation exists between the level of effort noted in NUREG/CR-2300 and that expected by the staff to conduct an IPE that is well in excess of 8,100 person-hours.

Response - PRA is a developing technology, and its efficiency is constantly being improved. It is therefore not appropriate to compare the resources needed to perform a PRA 5 or more years ago to the resources needed to perform the study today. A letter dated June 25, 1987, from Joseph Fragola, SAIC, to Themis Speis stated that: "Using computer work stations and supporting codes to integrate the input information, generate the event and fault trees, and to provide for configuration management, has offered substantial labor saving."

8.3   Training uninitiated utility personnel on PRA technology is clearly not the answer in the time frame allowed by the Generic Letter. It takes about a year to train a good systems engineer as a risk analyst. Ignoring this reality could conceivably result in less-than-adequate IPEs conducted by insufficiently trained personnel.

Response - Utilities are not expected to train personnel as PRA experts. The Generic Letter expressed the belief, however, that the most benefit to the utility would result from development of a cadre of utility personnel with good understanding of the IPE models and implications of the IPE conclusions as far as the plant design and operations are concerned.

8.4   What does the staff perceive the level of effort to be for the back-end analysis? What would be the appropriate partitioning of the front-end to the back-end?

Response - The estimate of 8,100 person-hours referred to in the submittal guidance document includes both the front-end and the back-end. The staff believes that most of the needed information on the back-end is presently available. The staff estimates the back-end effort somewhere between 1 and 2 person-years.

The features unique to any plant are not the phenomenological behaviors but are components, systems, and configurations within the plant itself. The dominant portion of the IPE effort would therefore appear to be in the front-end. It is where the greatest expectation reasonably exists to discover vulnerabilities that may be rather readily corrected.

9.   TREATMENT OF HUMAN FACTORS

9.1   The staff asked for a very specific discussion of human recovery action. I would like to know more clearly what the staff means by human recovery actions. In past studies, we've seen human recovery refer to all kinds of things. Sometimes at the most trivial level it's simply a manual backup to a failed automatic system. Other times it is repairing or restoring initially unavailable or failed equipment. Sometimes it is operator action involved in the EOP and executing the EOPs. Sometimes it's doing some action for which there aren't procedures written.

Response - Whether recovery action has a written procedure or not, if the action is important to the plant response, then the action should be described. Unless proper justification is provided, all important recovery actions should have written procedures.

9.2   Should "recovery action" be understood to mean any operator action?

Response - The term "recovery action" should include any operator action that the analysis would show is significant for the plant's ability to respond to an accident.

9.3   Can the staff clarify the intent of what operator actions are supposed to be modeled in the back-end, and which modeling errors of commission would be breaking new ground?

Response - The intent was not to use the IPE process to start a new approach. The staff's intention is not to "break new ground" in modeling errors of commission, but, in the recovery process, the analyst must take into account the information available to the operators and the procedures available to the operators as a contributor to not taking proper action.

9.4   A number of utilities may credit staff actions based on plant knowledge in lieu of specific procedures. Rather than rejecting nonproceduralized actions out of hand, the staff should be willing to review these recovery actions on their own merits.

Response - The analyst's judgment should be reflected at that point. The staff, however, expects that all assumed or modeled recovery actions will have written procedures. Most often the staff has received justifications for the assumptions of success for nonproceduralized actions based solely on time available for such actions. The staff does not believe this type of argument to be correct. There is much to be gained by pre-planning.

9.5   Should utilities infer that the list of Human Reliability Analysis (HRA) methods presented at the IPE workshop are acceptable for use in the IPE?

Response - HRA methodology is not mature and therefore will not allow the establishment of specific acceptance criteria on HRA methods. The best guidance available can be found in the staff's review of past PRAs. Additional references on treating human factors have been added to this document.

The objective of doing the analysis is not to establish the process of doing a human reliability analysis, but to make the plant safer through the human reliability analysis and subsequent accident management program.

9.6   Will there be HRA guidance in the final form of NUREG-1335?

Response - The staff is not issuing guidance on human factors in the document. Utilities should use their best judgment while keeping in mind the current state of technology. In the past, some PRAs have used common sense, good methods, and good approaches in treating human factors. It is recognized that the technology is still evolving, and the best the staff can offer is the status of that technology. Additional references, however, have been added.

9.7   Would an acceptable approach be one where 20 to 30 major cognitive operator actions were first identified and then a sensitivity study or importance calculation performed?

Response - Twenty actions may not be adequate, although screening and then conducting a sensitivity study is a very sound approach.

9.8   Draft NUREG-1335 states that sequences are to be listed where human error is less than one in ten. There are hundreds if not thousands of such human errors that are less than one in ten. In addition, PRAs implicitly exclude certain human errors that we know are very unlikely or will not be in the dominant accident sequences. There appears to be the need for additional clarification in terms of what is expected of utilities.

Response - Important action types might range from manual verification of automatic actions, execution of EOPs, and restoring unavailable systems to repairing failed components. Low values of human error rates depend upon the type of recovery actions required, e.g., an error rate of 0.001 per demand would be low if manual verification of automatic action is required, while an error rate of 0.1 per demand would be low if there were little time to act or procedures were not available for the required recovery action.

The screening of human actions by putting in high failure rates for the human action on an initial evaluation, and subsequently identifying the leading sequences, is a process that should cull out many of the unimportant failures. The rest of the analysis can then focus on the more important failures. The text has been revised, however, with regard to the one in ten listing of human actions.

9.9   Reference to the screening criteria in Appendix 2 of the Generic
         Letter, in combination with low human error rates in recovery
         actions, is inappropriate and should be deleted.

Response - Without proper justification, it is inappropriate to exclude
dominant or otherwise important accident sequences because of low human error
rates. These sequences should be exposed so they can be viewed and placed in
their proper perspective with respect to the rest of the IPE and understood
within the framework of the accident management program.

10. DATA, UNCERTAINTY, AND TREATMENT OF COMMON-CAUSE FAILURE

10.1   It is understood that plant-specific data were to be used to calculate
         certain system failure rates. How do we get around the problem where
         we might have a system that has had relatively infrequent failures
         such that we cannot draw statistically meaningful conclusions?

Response - Utilities should use plant-specific data only when statistically
meaningful data exist. Otherwise, generic data should be used along with the
rationale for using the generic data.

10.2   The common-cause failure methods put forth in NUREG/CR-4780 (Ref. C.5)
         are good, but require a large effort on somebody's part (NRC or EPRI)
         to generate a good common-cause failure data base. If analysts use
         the common-cause beta factors out of NUREG/CR-4780, they are going to
         dominate all the answers.

Response - The common-cause failure rate data base as it exists today is sparse.
Although the methodology is good, the data base needs to be improved with time.
The analyst cannot ignore the potential for common-cause failures, however, but
must look at the contribution to the data base and apply the beta factors or a
similar parametric device in a manner that makes engineering sense. Past PRAs
have set precedent in a reasonable way.

10.3   The proper characterization of uncertainty is key to understanding
         probabilistic results. The less one knows, the more important it is
         to quantify the uncertainty. Uncertainty quantification lets the
         analyst communicate his confidence or state of knowledge about the
         study.

Response - The staff agrees with this comment.

11. IPE DOCUMENTATION AND SUBMITTAL FORMAT

11.1   NUREG-1335 states: "It is not necessary to submit all of the documenta-
         tion needed for a review. What is existing should be cited and it
         should be available in usable form." Should this be interpreted to
         mean that a summary document should be submitted in which each of the
         items listed in the NUREG-1335 document is addressed by citation to
         documentation that exists at the plant; or does the staff visualize a
         several-volume PRA-like submittal?

Response - The staff does not expect or anticipate a five-volume risk analysis treatise from each utility. In essence, the submittal document should: (1) Be a reasonably complete summary of the effort, (2) indicate how the process was done, and (3) allow judgment to be passed on how well the process was done and how well it measures up to the specific objectives identified in the Generic Letter and introductory section of this document.

The submittal should be somewhat closer to a summary document, but certainly not an abstract because it needs to have enough substance for an adequate review.

11.2   What level of detail for each item in the standard table of contents is to be in the summary report to be submitted to the NRC, as opposed to what is to be retained by the licensee as backup information?

Response - The level of detail should be sufficient to enable the NRC to understand and review the validity of the results and conclusions of the IPE and to pass judgment as to whether or not the IPE has met the Generic Letter 88-20 objectives. Some submittals may require more detail than others in order to address certain unique plant-specific features. When in doubt, additional detail should be provided in support of the findings and thereby prevent a series of requests for additional information.

11.3   For those having performed and documented all or a majority of their PRA or IPEM, complete rearrangement of an established document in order to fit a standard format is unwarranted. The only requirements may be, as stated by the staff at the IPE workshop, that the utilities determine that the [previously completed] PRAs reflect plant configuration as of a given date.

Response - For sites that have completed (or nearly completed) a PRA prior to the IPE initiation date, conformance to the NRC standard format contained in Table 2.1 may be unnecessary and place an undue burden on the utilities' resources. (For such specific cases, justification for a different format should be provided to the staff along with the IPE submittal plans.) As a minimum, a "road map" in the form of a short document in the standard format of Table 2.1 with sections referenced to the existing analysis should be provided, along with the existing analysis. The staff will review and respond to such plans on a case-by-case basis.

11.4   What level of detail is required in response to the Generic Letter? Could it be a five-page response or a one-page response or a ten-page response?

Response - The response need not be extensive but should provide a clear identification of the IPE option chosen, the particular plans, schedules, and milestones. This could be provided in a few pages.

11.5   Submission of all system descriptions, fluid system simplified diagrams, electrical diagrams, and fault trees will result in a very large volume submittal. If this is not the intent, what specifically does the staff want licensees to submit with respect to system logic models and associated reference information?

Response - Only the front-line and support system descriptions and simplified diagrams considered in the IPE are to be submitted. All fault tree diagrams should be retained by the utility and be readily available upon request. The fault trees will be reviewed and audited on a case-by-case basis and need not be included as part of the IPE submittal.

12. VULNERABILITIES AND TREATMENT

12.1 Could the staff address the meaning of vulnerability treatment? It could mean different things to different people and have very broad scope.

Response - The utility should decide if it has identified a specific vulnerability, or weakness, and whether or not some corrective action is needed. The staff may also look at vulnerabilities for which no fixes were proposed or where potential vulnerabilities were not identified by the licensee.

Sequences that meet the screening criteria may need to be expanded further in order to understand why these sequences are above the screening criteria. If a weakness is found, the utility may decide it is a vulnerability and propose a fix. The screening criteria by themselves, however, do not define a vulnerability.

Examples of plant features and operator action identified as being important for either preventing or mitigating severe accidents can be found in NUREG/CR-4920 (Ref. C.6). Without specified criteria, vulnerabilities were identified that could, for example, result in suppression pool bypass or early containment failure. This information can be used to examine the subject plant and determine if the same or similar plant features and operator actions will be of value in reducing the significance of identified vulnerabilities.

12.2 If a utility identifies a vulnerability and proposes some type of fix, why must the utility also list all the different options considered and the pros and cons of those options?

Response - All strategies considered for implementation to correct outliers need not be included in the final report. Rather, just those corrective actions selected for implementation need be described. If all the alternatives have been dropped from further consideration because of high cost, it is important to discuss how less expensive alternatives were sought. The submittal should contain sufficient discussion so that a reviewer can be confident that a reasonable effort to address each identified vulnerability has been performed, whether or not a fix has been implemented.

12.3 Can part of the resolution of vulnerabilities come out of the accident management program? Accident management criteria have not been issued although utilities may want to use part of the accident management program and plan.

Response - The intent of the IPE program is to look at the plant, with its configuration as it exists, from the point of potential severe accident vulnerabilities and try to determine whether there are fixes of reasonable cost that would reduce identified vulnerabilities. There may be some things that are uncovered in the IPE that may be handled best in a procedural fashion. These would provide the "flesh on the bones" of the accident management program. The staff does not expect that the accident management program would involve hardware changes although they cannot be entirely discounted.

12.4   It is more appropriate for a utility to "identify" vulnerabilities than "define" them.

Response - The staff is interested in understanding the criteria used to identify vulnerabilities. The word "define" appears to be more appropriate in that context because the reporting process must go beyond simple identification.

13.   CONSIDERATION OF UNRESOLVED SAFETY ISSUES AND GENERIC SAFETY ISSUES

13.1   A number of unresolved safety issues (USIs) and generic safety issues (GSIs) have been resolved via rulemaking or issuance of Generic Letters. Could the staff provide a list of those USIs and GSIs that have not been resolved so that utilities could address them within the IPE framework?

Response - One beneficial element of the IPE focus is to allow resolution of the unresolved generic issues. The full listing of generic issues appears in NUREG-0933 (Ref. C.7), which is updated annually.

13.2   For plants with existing PRAs or IPEs, it may be more efficient to resolve A-45 separately; this option should be left open to licensees.

Response - USI A-45 does not exist as it has been subsumed into the IPE. This option would reopen the A-45 issue and is not acceptable.

13.3   Specific guidance is required regarding the scope of the analysis needed to demonstrate the adequacy of the decay heat removal capability of a plant. This guidance should include specification of what constitutes acceptable capability versus unacceptable capability.

Response - Six case studies were performed under USI A-45, Decay Heat Removal (DHR) Requirements. The purpose of these studies was to identify potential vulnerabilities in the DHR system, to suggest possible modifications to improve the DHR capability, and to assess the value and impact of the most promising alternatives to the existing DHR system. These studies identified potential vulnerabilities and corrective actions without prescriptive acceptance criteria.

Insights from these studies can be used during the evaluation of licensee DHR systems. The references are as follows:

1. "Shutdown Decay Heat Removal Analysis of a Westinghouse 2-Loop Pressurized Water Reactor. Case Study," NUREG/CR-4458, SAND85-2496, Sandia National Laboratories, March 1987.

2. "Shutdown Decay Heat Removal Analysis of a Westinghouse 3-Loop Pressurized Water Reactor. Case Study," NUREG/CR-4762, SAND86-2377, Sandia National Laboratories, March 1987.

3. "Shutdown Decay Heat Removal Analysis of a Babcock and Wilcox Pressurized Water Reactor. Case Study," NUREG/CR-4713, SAND86-1832, Sandia National Laboratories, March 1987.

4. "Shutdown Decay Heat Removal Analysis of a Combustion Engineering 2-Loop Pressurized Water Reactor. Case Study," NUREG/CR-4710, SAND86-1797, Sandia National Laboratories, July 1987.

5. "Shutdown Decay Heat Removal Analysis of a General Electric BWR3/Mark I. Case Study," NUREG/CR-4448, SAND85-2373, Sandia National Laboratories, March 1987.

6. "Shutdown Decay Heat Removal Analysis of a General Electric BWR4/Mark I. Case Study," NUREG/CR-4767, SAND86-2419, Sandia National Laboratories, July 1987.

13.4   The staff stated that USI A-45 is to be enveloped within the IPE although other programs are ongoing. What is the staff expecting now, and what is the staff expecting later?

Previous staff analysis indicated that decay heat removal vulnerabilities were likely to be plant specific, and one universal fix would not be cost effective. Had the Commission decided not to move forward with the IPE program, the staff would have recommended that each plant do a vulnerability search of its decay heat removal system. The IPE will accomplish this task and therefore A-45 is resolved.

With regard to external events, staff guidance will ultimately address this aspect of A-45 as well.

13.5   What additional analyses and documentation are required beyond that dictated by other IPE requirements to support [USI/GSI] resolution?

Response - Initiation of staff review of a USI or GSI requires all of the following:

• The methodology should be capable of identifying vulnerabilities associated with the USI and GSI being addressed.

• The contribution of each USI and GSI to core damage frequency or unusually poor containment performance are identified and quantified.

• A description of the technical basis for resolving any USI or GSI is given.

14. NRC STAFF REVIEW AND REVIEW GUIDANCE

14.1   Why has the IPE review guidance document been dropped prior to the IPE workshop?

Response - The present expectations and intent is that the review guidance be procedural guidance for use within the staff. It is not the intent to establish new guidance for the process of doing the IPE or to set forth previously unheard of acceptance criteria. It is primarily to be directed toward procedural guidance on how the staff will handle the process and carry out a review of IPE information submitted in accordance with the submittal guidance.

14.2   There is concern that whatever is in the review guidance document could affect how one does the IPE. The review guidance should be furnished to the utilities as soon as possible to when NUREG-1335 is issued and the clock starts.

Response - Review guidance has been included as Appendix D to this document.

14.3   Following a favorable IPE report, would it be acceptable to have the report referenced in future licensing analysis?

Response - It is possible to reference the IPE report in licensing analysis although it is not the staff's primary purpose. It should be recognized that the IPE review may not necessarily be the same kind of review that would normally occur in the licensing process.

14.4   The staff should discuss in NUREG-1335 how ratcheting will be avoided while including the clearinghouse aspect.

Response - The IPE submittal will be judged against the objectives stated in Generic Letter 88-20. The utilities should keep this in mind when putting together their IPEs. With regard to the clearinghouse aspect, the intent is to make available to the utilities whatever interesting insights or information results from the IPE reviews. The utilities should decide for themselves how relevant the information is with respect to their plants.

14.5   Does the staff see the IPE review information as necessarily requiring an amendment to the utilities' IPE effort?

Response - The intent was not to ratchet the utilities but rather to act as an information clearinghouse because of the unique position that the staff would be in when all this IPE information became available. Each individual utility will have to decide, based on their own judgment and information available, whether corrective action is warranted.

14.6   Is there the possibility of a two-step process in which utilities could submit preliminary results and have them reviewed by the staff prior to spending resources on treating vulnerabilities?

Response - Utilities should submit their best efforts as a final IPE document and not submit an interim report for staff review. However, the staff will consider requests from utilities for discussion of specific issues during the IPE process and will accommodate such requests whenever possible.

14.7   Will the IPE reviews be conducted primarily by RES or NRR?

Response - The IPE review effort will be a combined effort between the NRR staff and RES staff although submittals will be made solely to the NRR staff.

14.8   Draft NUREG-1335 should be clarified to indicate that additional staff reviews of previously submitted PRAs for IPE compliance will not be necessary except where additional submittals are involved.

Response - The NRC staff will review the IPE (PRA) submittal and determine if the licensee has met the intent of the Severe Accident Policy Statement, i.e., the objectives of Generic Letter 88-20. The purpose of the IPE staff review is therefore different from reviews previously performed on past PRA submittals. Although it is likely that an IPE review of a previously submitted PRA would be less intensive, a review would be nevertheless required.

A licensee submittal must specifically address the information requested in Generic Letter 88-20. This submittal may reference a prior PRA, but the specific questions posed in the Generic Letter must be addressed in a separate response. Utilities that choose to use an existing PRA, NUREG-1150 analyses, or similar analyses (IDCOR test application) should:

1.   Certify that the IPE meets the intent of and responds to the information requirements of the Generic Letter, particularly with respect to utility involvement,

2.   Certify that it reflects the current plant design and operation, and

3.   Submit the results on a schedule shorter than 3 years.

A dependency matrix should also be included.

14.9   As guidance and review of the IPE process evolve, the potential exists for guidance to differ from that specified in the Generic Letter and NUREG-1335. When this occurs, the Generic Letter and NUREG-1335 must be considered the final authority, unless specifically documented otherwise by the staff.

Response - The staff agrees with this comment.

14.10   If one is to perform a PRA, it is our understanding that the staff is willing to accept the analyst's judgment of what is appropriate within the guidelines in any of the cited NUREGs (NUREG/CR-2300 (Ref. C.4), NUREG/CR-2815 (Ref. C.8), and NUREG/CR-4550 (Refs. C.9 through C.15)).

The staff agrees with the intent of this statement, with the additional observation that more current documents contain updated data or methodologies. The more current documents should be given primary consideration. There may be instances, however, where the staff may make inquiries into the basis for the analyst's judgment upon review of the IPE submittal.

15. INDEPENDENT REVIEW OF THE IPE

15.1   Utilities that do not have PRA expertise already in-house will have to train people. Training can only be done effectively if the people involved participate in the actual PRA or IPE process themselves. After those people have been involved, they are no longer independent, so how does one satisfy the staff's independence requirements once someone has been trained or participated in the PRA itself?

Response - The staff recognizes that licensee organizations, and in-house expertise in the area of probabilistic analysis, are quite variable. The emphasis here is on the independence of review from the conduct of the analysis for purposes of quality assurance. The staff expects, of course, that all utilities have the most expert knowledge in-house of their own plant, systems configurations, and operating practices and procedures.

15.2   Can the staff explain how the statement in the Generic Letter, "This independent in-house review is to validate both the IPE process and its result," is to be carried out? If the IDCOR IPEM methodology is used, one would have to do a PRA to validate it or vice versa.

Response - The term "validate" is to mean an in-house critical review of the IPE such that considerable confidence in the results and conclusions can be gained.

15.3   It is essential that studies as important as these IPEs be subjected to at least some outside, independent review.

Response - For some IPEs, it might be prudent to have an outside contractor review the IPE submittal prior to submittal to the NRC staff. Such a review could provide useful feedback from sources independent of and unattached to the utility being examined. Review by an outside party, however, is not a prerequisite for meeting the IPE objectives and therefore is not to be an explicit requirement of the IPE.

16. EQUIPMENT SURVIVABILITY

16.1   The staff is urged to apply a test of reasonableness regarding credit taken for equipment used in severe accident response in lieu of rigid qualification records.

Response - The staff agrees with the implied conclusion that formal environmental qualification requirements are not applicable to the IPE and accident management process. When credit is taken for equipment in severe accidents, an assessment should be made of the ability of the equipment to perform the function for a specific period of time considering exposure to temperature, pressure, aerosol loading, radiation, and moisture. The degree of credit should be dependent upon some evidence of the capacity of the equipment to survive or operate in the expected circumstances for the accident sequence, in addition to the impact of radiation or other adverse conditions on personnel needed to operate such equipment.

16.2   Does the staff believe that there is enough information available on equipment qualification in severe accidents and therefore utilities do not need to do any plant-specific analysis?

Response - Utilities are going to have to justify the use of equipment and the conditions under which they are exposed in order to take credit for the equipment in severe accidents. The staff is not looking for a prescriptive analysis that shows a direct tie with experiments, but rather a common sense approach to showing that this piece of equipment can be expected to work under severe accident conditions. The only information available is the standard equipment qualification information that comes in the licensing process, which is not very good from a reliability standpoint because it does not give a large number of data points.

The staff is not suggesting that equipment be qualified for severe accidents under 10 CFR 50.49 requirements. If the data do not cover the range of conditions expected during a severe accident, then the data would presumably be extrapolated. Use engineering judgment. Additional guidance can be found in NUREG/CR-5313, Equipment Qualification Scoping Study (Ref. C.16).

16.3   I think the staff should be very specific about equipment qualification and how 10 CFR 50.49 does not apply to severe accidents.

Response - The staff agrees and has made a special point of putting it that way in the text.

16.4   With regard to the equipment survivability issue, would you describe the judgment used in WASH-1400 for the turbine-driven pumps in a steam environment (where a failure rate adjustment was made) as the kind of judgment to be used in the IPE?

Response - That is a good example. If a pump is operating in an environment 10 to 20 degrees higher than its qualified condition, the analyst may want to adjust the failure rate. The analyst should look at the clearances and what would be expected on increasing temperature. For example, will the lubricant break down at high temperature? Will the seals be gone? The analyst will have to apply good engineering sense.

17. STAFF RESPONSE TO IPE SUBMITTALS

17.1   How does the staff intend to respond to the submittal plans?

Response - There will be a written acceptance of the submittal plans. The submittals will be entered on the docket record. NRC responses (i.e., NRC's acceptance of the submittal plans) to the utilities are expected to be made within 30 to 50 days after the plans are submitted.

17.2   How does the staff intend to respond to the IPE submittals?

Response - As discussed in Appendix D, the staff will use an IPE Evaluation Report documenting the review process and the conclusions relative to the objectives stated in Section 1.1.

17.3   It is the staff's intent to hold discussions with utilities, answer questions, clarify guidance, etc. Is there some mechanism, however, that would allow other utilities to know the kind of clarification guidance the staff is giving an individual utility, or owners' group, or NUMARC?

Response - An official meeting between the staff and a utility will result in a written summary being placed in the Public Document Room. The summary would therefore be available to the public. Should a utility identify something that could be of interest to all utilities, that utility might want to share it with NUMARC. NUMARC could then determine its generic implications. Based on such a determination, the staff would consider having a meeting that would deal with that issue or a set of issues. In other cases, individual utilities can come to the staff with their unique questions related to the IPE performance, and these will be discussed on a case-by-case basis.

18.   EMERGENCY OPERATING PROCEDURES (EOPs)

18.1   There is the implication in NUREG-1335 that EOPs are needed for operator actions in the containment event tree. Is the implication correct, or are there different criteria for operator actions in the back-end analysis and in the front-end analysis?

Response - The operator need not be in an EOP for assurance that he will perform a specific action. There are certain actions that might be needed 3 days later, for example, that are not necessarily in the EOP. If it is an important action for mitigating a sequence, the action should be well thought out and available to the operator although the operator may not necessarily be in an EOP. (See also the response to 9.4.)

18.2   Would the staff consider an action in the EOP for preventing core damage that (although not carried out in time to prevent core damage) may prevent vessel failure? Could credit be taken in a similar manner for the second case?

Response - Given proper justification and consideration of the usual concerns of human error and equipment failure, credit may be taken for the second case.

18.3   Considering the past 10 years of interactions with the staff on developing EPGs and in implementing EOPs and EOIs, does the staff envision another extensive iteration of that type with regard to severe accidents?

Response - It is the responsibility of each utility to ensure that procedures for which it takes credit in the IPE are in place and that operators have been trained on them. We do not see a need for generically extending the EOPs at this time.

19. ACCIDENT MANAGEMENT

19.1   If a BWR has implemented EPG Rev. 4, where does it go from there? Would there be little else to do under accident management?

Response - Implementation of Rev. 4 of the EPG is certainly a very important aspect of accident management but not the whole answer. Other aspects of the program involve: additional procedures to deal with important sequences and equipment failures identified through the IPE; training for severe accidents for licensed operators, technical support staff, and key managers in the licensee's emergency response organization; guidance and computational aids for the technical support staff; evaluation of information needs and availability during severe accidents; and evaluation of the licensee's decisionmaking process for severe accidents.

With regard to procedures, the IPE study performed for each plant will provide a great deal of technical information on which further enhancements to utility accident management capabilities would be based. Another source of information will be a set of generic accident management strategies or "PRA lessons-learned" to be compiled by the NRC and provided to industry. The expectation is that additional accident management procedures or guidance will be implemented by utilities to reflect the insights obtained through the IPE and the utility's evaluation of the NRC accident management strategies.

19.2   It is requested that the Commission consider suspending the implementation of Regulatory Guide 1.97 (Ref. C.17) that is currently in progress for many utilities with an eye toward spending those funds in a cost-efficient manner on the instrumentation likely to be implemented as a result of the accident management program. The backfit cost on instrumentation is extremely expensive, and, if the funds could be allocated more on sophisticated requirements and less in the way of deterministic requirements, it would be a major benefit to the utilities.

Response - It would be a mistake to suspend work on Regulatory Guide 1.97. It is not the intent of the accident management program to overturn Regulatory Guide 1.97, nor is it the intent of the program to require major modifications to instrumentation. A more balanced approach is to go through the scenarios, find the severe accident vulnerabilities, implement proper procedures, inform the technical support people of the kinds of accidents that they should be looking at, and understand whether or not the information from a specific instrument will be available when needed. If a piece of information is not available when needed, then it would be time to rethink the procedure or make a modification to the instrument. If the instrumentation is going to be available under Regulatory Guide 1.97, that would be acceptable.

19.3   It would be difficult to ensure that the operating staff can make
        correct or best decisions for public safety without having direct
        information about the radioactivity that is being or can be released.
        Would the staff comment on how best a utility should proceed in light
        of the uncertainties and knowledge that is going to be required?
        This also reflects the need to address radioactivity as part of the
        IPE.

Response - The staff is continuing to work with industry, through NUMARC, to
better define types of guidance and computational aids that should be provided
to the emergency response teams. The accident management guidelines under
development by NUMARC are expected to provide further guidance to utilities on
this topic. It is anticipated that these guidelines will be available in late
1990.

It was not the intent of the staff to omit information related to radiation
levels from the list of information that should be made available to the
technical support staff. The list was only intended to give examples of the
types of information that the staff would want to have made available. A more
complete list would certainly include information related to radiation levels
inside containment as well as in areas to which recovery teams may need to
have access.

20. OPERATOR TRAINING

20.1   In the past, INPO has played a key role in training plant staff. Has
        the staff considered INPO within the accident management framework?
        In particular, in the operator training area, there was an actual
        Commission policy statement that deferred it to INPO.

Response - The staff will be working with NUMARC to define what is appropriate
coverage of severe accidents in utility training programs and to define a
mechanism for implementing and evaluating such programs. It is our intent to
involve INPO in these interactions. The staff recognizes the commitment
regarding INPO accreditation of utility training programs for licensed operators
and considers the INPO training accreditation process as a possible means of
ensuring adequate severe accident training for technical support staff and key
managers in the utility emergency response organization, as well as for licensed
operators. Note, however, that INPO has not (up to this time) come forward
with a program.

20.2   There appears to be a contradiction with regard to emergency response
        onsite versus offsite. There appears to be negative "onsite"
        training because the operator is stopped before he can solve the
        problem in order to have offsite people do their thing.

Response - Traditionally, the emphasis in annual emergency preparedness
exercises has been on evaluating the effectiveness of offsite response.
Significantly less importance has been placed on the ability of the utility
staff to effectively prevent core damage and mitigate offsite releases. For
example, numerous options for averting or arresting core damage are usually
identified by the utility staff during an emergency response exercise, but

these measures do not generally receive a detailed technical assessment by the utility, particularly with regard to whether the proposed fixes would actually have been effective.

An important objective of the NRC accident management program is to have utilities better exercise those aspects of emergency operations related to the prevention and mitigation of severe accidents and to increase the staff's emphasis on this area as part of ongoing regulatory activities. An increased emphasis on onsite response (accident prevention/mitigation) during the off-year emergency preparedness exercise (i.e., the small-scale exercise held every other year, typically without the full participation of State and local governments) is one approach that will be pursued in this program. NRC Information Notice 87-54 (Ref. C.18), which reminds utilities of the flexibility of the emergency preparedness rules in this regard, is a first step toward this goal. With regard to increasing the staff's review efforts in the area of accident management, a number of changes to present practice will be considered. These include placing a greater emphasis on the technical adequacy of preventive and mitigative measures identified by licensees during annual emergency preparedness exercises and periodically conducting detailed assessments of accident management capabilities during annual exercises.

20.3   Presently the lines of authority in the control room are very clearly defined. Do you envision that the technical support staff and other managers that are going to be trained in accident management are going to have some sort of a qualification as a "severe accident manager" and that there would be some point in an accident at which they would usurp the shift supervisor's authority?

Response - The staff envisions neither major changes to the lines of authority established by licensees in response to existing regulation and guidance nor new requirements that technical support staff and managers be qualified for accident management. Rather, the focus of accident management is on changing the thinking and the planning process so that utilities can more effectively deal with accidents beyond the scope of the existing emergency operating procedures. Two important aspects of this effort will involve (1) incremental improvements to the emergency operating procedures to better deal with potential severe accidents, and (2) increased training for technical support staff and key managers on severe accident insights and accident management strategies.

20.4   Does the staff expect that the accident management program or severe accident mitigation will become part of the operator licensing procedure?

Response - The staff will be working with NUMARC to define a mechanism for implementing and evaluating severe accident training programs. We intend to pursue the INPO training accreditation process as a possible means of ensuring adequate severe accident training. To the extent that training for severe accidents receives additional attention in the INPO program for licensed operators, there would be a link between severe accidents and operator licensing.

21. ACCIDENT STRATEGIES

21.1   Is the staff going to put out a vulnerability list? And if so, does the staff expect that there will be documentation by each of the utilities as to how they respond to the staff's list?

Response - The staff will not put out a "vulnerability list" but will issue a list of generic accident management strategies. The list is to contain certain generic accident management strategies identified by NRC on the basis of existing PRAs. They are strategies and not procedures. They point to general types of actions that utilities might want to consider for inclusion in their procedures, either in the emergency operating procedures or special procedures, for example, for the Technical Support Center. The staff intends to issue generic strategies with the intent that utilities would consider them as they are doing their IPEs and are learning about the risk aspect of their plants. Any requirements regarding utility evaluation of the accident management strategies and documentation of the results of these evaluations will be clarified when the list is sent to the utilities.

21.2   It is important that the staff (1) provide better guidance on what the utilities are to do with accident management strategies, and (2) specify how the utilities are expected to respond to the Generic Letter Supplement and how utilities are to keep this documentation.

Response - Any requirements regarding utility evaluation of the accident management strategies and documentation of the results of these evaluations will be clarified.

21.3   What is the format of the accident management strategies that the staff will send out in the near future?

Response - At the moment, the format of the accident management strategies is simply a subject title list, as presented in SECY-89-012 (Ref. C.19). The letter that will formally transmit the strategies will include a more complete description of each proposed strategy and a technical assessment of each of the accident management strategies to ensure, to the extent possible, that the strategies will not detract from overall safety. The letter will also provide evaluation guidance and cautions for each strategy to provide added assurance that use of the strategy will not detract from safety. Of course, the positive and negative impacts of each strategy may be different for each plant. Consequently, the evaluation of the feasibility and effectiveness of each strategy should be performed by individual licensees.

22. APPLICATION OF 10 CFR 50.59 CRITERIA TO SEVERE ACCIDENTS

22.1   If a utility wants to make a procedure change or hardware change that relates to severe accidents, how does it satisfy the 10 CFR 50.59 criteria? The 10 CFR 50.59 regulation requires evaluation of the changes against the accidents described in the FSAR. The FSAR does not consider severe accidents.

Response - The NRC Working Group on 10 CFR 50.59 is actively working to develop guidelines for conducting safety evaluations in accordance with 10 CFR 50.59.

An early draft guidance document, prepared by industry, was reviewed by the NRC Working Group, and staff comments on that draft were provided to NUMARC/NSAC in a 1988 letter to Mr. Thomas E. Tipton (Ref. C.20). NUMARC/NSAC revised the guidance document in response to comments from utilities, other industry organizations, and the NRC. NUMARC's proposed "final draft" was received in mid-November 1988. The "final draft" was widely distributed within the NRC for comment. The NRC Working Group has met to discuss the comments received and issues involved, e.g., what constitutes a reduction in the margin of safety or an increase in the probability or consequences of an accident? The NRC Working Group is preparing proposed staff positions on these and other issues. Additional information may be found in a May 10, 1989 letter to Mr. Tipton (Ref. C.20).

23. INTEGRATED SAFETY ASSESSMENT

23.1   Under ISA, should it be understood that no license amendment is required for the process, or for the modifications and the ranking of the modifications, or both?

Response - No license amendment is required to either participate in the process or the ranking.

23.2   There is the belief that any money spent unwisely reduces safety because less money is then available to resolve safety issues. Is it the intent of the ISA program to allow utilities to increase plant reliability and, in effect, free money for safety changes; or is it only to rank safety issues?

Response - The intent of the ISA program is to allow utilities to rank and put on an integrated schedule all issues, not only NRC safety issues, but those issues that the utility feels are important for perhaps not merely safety reasons but other reasons as well.

23.3   Is the staff saying that the IDCOR IPEM, with or without the enhancements, is unacceptable for the ISA, although it is acceptable for the IPE?

Response - The IDCOR IPEM without the staff enhancements should not be used for the IPE analyses, nor is it acceptable for the ISA. With regard to the ISA, the staff will pass judgment as to the adequacy of the IPEM submittal with enhancements on a case-by-case basis.

23.4   Statistically significant plant-specific information would be expected to be used to perform and satisfy the IPE. Does the ISA require an expansion of these important components?

Response - The ISA is simply looking for the use of plant-specific failure rate data if applicable generic data are not available.

23.5   Can the ISA be applied to multiple-unit sites, for example, in setting prioritization at the site, or is it to be applied only to one plant at a time, i.e., plant-specific rather than site-specific?

Response - With proper justification, utilities could combine schedules and account for differences between units. That is the utility's option, too. It is important to note that in an ISA utilities will be dealing with licensing project management and the people they normally deal with otherwise. NRC questions can be worked out the way questions are normally worked out in the licensing process.

23.6   Does the ISA option exist for those plants that have insufficient plant-specific data? For example, could those plants use generic data in the PRA in combination with a program to collect operating data in a form suitable for future updates?

Response - Plants using generic data in their PRAs where no plant-specific data exist can choose the ISA option.

23.7   The PRA can be used in the license renewal process, risk management area, which is the same thing as ranking components and systems, and so forth. Is it true that the PRA does what ISA does? Are they synonymous?

Response - No. The key that separates the ISA option from just doing the IPE is the ability to do the integrated scheduling on the basis of your best estimate of the risk of the plant. The ISA program is a process that uses PRA as a decision tool.

23.8   After a utility does a PRA, could it not also choose to do integrated scheduling and not call it ISA?

Response - The NRC is simply offering a more formalized process for going through the PRA and working with the utility in establishing the ranking of various issues that would be identified, effectively establishing and formalizing the integrated schedule.

23.9   It is not emphasized enough that the sequences derived through the IPE process should be understandable at the RO/SRO level.

Response - Casting the sequence information in a form understandable by reactor operators and senior reactor operators may be a highly suitable method of transferring information from the IPE analysts to other parts of the utility's operations. This should be performed at the discretion of the utility, and, therefore, does not appear as part of the submittal guidance.

23.10   The [IPE Submittal Guidance] document does not emphasize the importance of the success criteria being established by virtue of performing realistic (best estimate) thermal-hydraulic calculations. Without realistic success criteria, many cost-beneficial solutions for accident management purposes might be obscured and the wrong set of sequences chosen for the purposes of control room operating crew training.

Response - In order that the IPE achieve the Generic Letter 88-20 objectives, a study of the plant must be performed that is both complete and realistic. For some plants, plant-specific thermal-hydraulic calculations may be required; for others, a suitably similar available analysis may be used. In either case, uncertainties should be recognized and a range given that operators can consider during severe accidents.

23.11   Relative to the IPE implementation schedule, when would the NRC be
          receptive to the review of such action implementation schedules and
          the basis for elimination of actions, i.e., before the IPE submittal,
          after, or any time?

Response - The NRC staff would be receptive at any time.

24. REGIONAL INSPECTIONS

24.1   The staff should carefully consider what is meant by "oversight
          through routine inspection." Commitments under the severe accident
          program in writing become current regulation. These commitments may
          result in tech spec violations if, for example, Mode 1 systems have
          to be taken out of service to test a system related to severe
          accidents.

Response - There were many options considered for severe accidents, i.e.,
rulemaking, bulletin, Generic Letter under 10 CFR 50.54. The staff opted for a
more performance-oriented approach by asking the utilities to commit to
implementation of an accident management program along lines of guidance
developed by representatives in NUMARC and EPRI and to verify the performance
through inspection programs. There is the potential for conflict with other
commitments. The staff will have to work it out on a case-by-case basis.

24.2   What is the intent of the risk-based inspection guide, and are the
          NRC regional offices going to use the inspection guide to follow the
          development and implementation of the IPE?

Response - The intent of the risk-based inspection guide is to give the
resident inspectors some guidance concerning the most risk-significant aspects
of their plants and to try to make that risk-significant information as plant
specific as possible. The risk-based inspection guide is not in any way tied
to the severe accident resolution process. It is also not tied to the IPE
program and is strictly voluntary.

25. GENERAL COMMENTS AND QUESTIONS

25.1   Is the staff going to reference Level 1, Level 2, and Level 3 PRAs
          that they consider to be acceptable for assessing vulnerabilities and
          comparing studies?

Response - There are many outstanding PRAs and reports that summarize the
insights gained from performing these PRA studies (see Appendix B). These PRAs
and reports are in the open literature, and the utilities should be aware of
them. The staff, however, does not intend to specify one or two PRAs as models
that licensees must use as models of acceptability.

25.2   It is inappropriate to generally require "state of the art" enhancement of existing analyses.

Response - The staff expects that utilities will use "state-of-the-art" methodology and most recent data available in quantifying their IPEs. For existing analyses, utilities should be aware of the limitations and shortcomings in their PRAs and update them as appropriate. The staff does not expect that there will be a substantial effort to update past PRAs to conform with current state-of-the-art methodology.

25.3   Why is NUREG-1150 not cited as an example of a CET methodology?

Response - The revised (1989) NUREG-1150 (Ref. C.2) can be used as a reference. However, the CETs employed in NUREG-1150 are extensive. As discussed in Appendix A, such extensive trees may not be necessary.

25.4   Would a utility making a 10 CFR 50.54(f) report in accordance with the Generic Letter 88-20 screening criteria also be required to report in accordance with 10 CFR 50.73 [Licensee Event Report System]?

Response - In the event that a specific plant modification is needed to meet current regulations, then an LER should be filed.

25.5   The request to include an assessment of the penetration elastomer seal materials and their response to prolonged high temperatures is better suited for the review criteria or technical guidance. To better focus industry understanding of what is being presented, the technical and style/content guidance should be separated.

Response - Technical items of special interest have been pointed out in the text to assure the staff that they would be submitted and are available when the IPE is reviewed.

25.6   Will an IPE be deemed inadequate if insufficient utility personnel were involved in the initial development of the IPE models; e.g., in the case where the utility has already performed a PRA, which it desires to submit as the basis for its IPE, but the PRA was largely performed by contractor personnel?

Response - The IPE will only be deemed inadequate if it fails to achieve the objectives put forth in the Generic Letter. How the PRA meets these objectives should be fully discussed, including the bases for understanding the important insights and limitations of the IPE, as well as utility involvement.

25.7   The front-end analysis documented in the initial NUREG-1150 documents provided rationale to limit the scope of the model to a selected list of initiators and phenomena to be quantified. In the event a licensee's plant does not differ materially from the plant designs analyzed in NUREG-1150, can such arguments also be used to dismiss the same initiators and phenomena from further consideration?

Response - Yes, provided that similarity is shown and justified between the reference plant and plant under examination.

25.8   The IPE process seems open ended with no discernible endpoint.

Response - In general, probabilistic studies can have very broad scope and appear open ended. The objectives found in Generic Letter 88-20 and the guidance provided there and in this document are intended to focus the study and effectively bound the effort. It is then up to the people that know best, the IPE analysts and the appropriate licensee personnel, to complete the study and fulfill the objectives of the IPE Generic Letter.

The IPE process itself will terminate following a satisfactory NRC review.

REFERENCES FOR APPENDIX C

C.1 USNRC, "Integration Plan for Closure of Severe Accident Issues," SECY-88-147, dated May 25, 1988.

C.2 USNRC, "Severe Accident Risks: An Assessment for Five U.S. Nuclear Power Plants," NUREG-1150, Vols. 1 and 2, Second Draft for Peer Review, June 1989.

C.3. F. T. Harper et al., "Evaluation of Severe Accident Risks: Quantification of Major Input Parameters," Sandia National Laboratories, NUREG/CR-4551, Vol. 2, Draft Revision 1, SAND86-1309, to be published.*

C.4 J. W. Hickman, "PRA Procedures Guide: A Guide to the Performance of Probabilistic Risk Assessments for Nuclear Power Plants," American Nuclear Society and Institute of Electrical and Electronic Engineers, NUREG/CR-2300, Vols. 1 and 2, January 1983.

C.5 A. Mosleh et al., "Procedures for Treating Common Cause Failures in Safety and Reliability Studies. Procedural Framework and Examples," Pickard, Lowe and Garrick, Inc., NUREG/CR-4780, Vol. 1, EPRI NP-5613, January 1988.

C.6 Brookhaven National Laboratory, "Assessment of Severe Accident Prevention and Mitigation Features," NUREG/CR-4920, Vols. 1-5, BNL-NUREG-52070, July 1988.

C.7 R. Emrit et al., "A Prioritization of Generic Safety Issues," NUREG-0933, Supplement 8, November 1988.

C.8 M. McCann et al., "Probabilistic Safety Analysis Procedures Guide," Brookhaven National Laboratory, Revision 1 to NUREG/CR-2815, Vols. 1 and 2, August 1985.

C.9 D. M. Ericson, Jr., (Ed) et al., "Analysis of Core Damage Frequency: Methodology Guidelines," Sandia National Laboratories, NUREG/CR-4550, Vol. 1, Rev. 1, SAND86-2084, to be published.*

C.10 T. A. Wheeler et al., "Analysis of Core Damage Frequency from Internal Events: Expert Judgment Elicitation," Sandia National Laboratories, NUREG/CR-4550, Vol. 2, SAND86-2084, April 1989.

C.11 R. C. Bertucio and J. A. Julius, "Analysis of Core Damage Frequency: Surry Unit 1," Sandia National Laboratories, NUREG/CR-4550, Vol. 3, Rev. 1, SAND86-2084, to be published.*

C.12 A. M. Kolaczkowski et al., "Analysis of Core Damage Frequency: Peach Bottom Unit 2," Sandia National Laboratories, NUREG/CR-4550, Vol. 4, Rev. 1, SAND86-2084, to be published.*

C.13 R. C. Bertucio and S. R. Brown, "Analysis of Core Damage Frequency: Sequoyah Unit 1," Sandia National Laboratories, NUREG/CR-4550, Vol. 5, Rev. 1, SAND86-2084, to be published.*

*Available in the NRC Public Document Room, 2120 L Street NW., Washington, DC.

C.14 M. T. Drouin et al., "Analysis of Core Damage Frequency: Grand Gulf Unit 1," Sandia National Laboratories, NUREG/CR-4550, Vol. 6, Rev. 1, SAND86-2084, to be published.*

C.15 M. B. Sattison and K. W. Hall, "Analysis of Core Damage Frequency: Zion Unit 1," Idaho National Engineering Laboratory, NUREG/CR-4550, Vol. 7, Rev. 1, EGG-2554, to be published.*

C.16 L. D. Bustard et al., "EQ Risk Scoping Study," Sandia National Laboratories, NUREG/CR-5313, SAND88-3330, January 1989.

C.17 USNRC, Regulatory Guide 1.97, "Instrumentation for Light-Water-Cooled Nuclear Power Plants to Assess Plant and Environs Conditions During and Following an Accident."

C.18 USNRC, "Emergency Response Exercises," Information Notice 87-54, October 23, 1987.

C.19 USNRC, "Staff Plans for Accident Management Regulatory and Research Programs," SECY-89-012, January 18, 1989.

C.20 Letters from C. E. Rossi, NRC, to Thomas E. Tipton, NUMARC, on draft guidelines related to § 50.59 reviews, dated May 12, 1988, and May 10, 1989.

*Available in the NRC Public Document Room, 2120 L Street NW., Washington, DC.

APPENDIX D

STAFF REVIEW GUIDANCE

In this appendix, the procedure by which the staff will review the IPE submittals is discussed.

The purpose of the staff review is to determine whether or not the IPE process was adequate to meet the four specific objectives listed in Section 1.1. If it has been determined that the process has met the objectives, the presumption will be that the examination of the plant has met the expectations of the Severe Accident Policy Statement.

In general, the staff is expected to perform an audit review of each submittal. The staff plans to review IPE submittals on a team rather than an individual basis and expects to use contractor personnel as part of each team. However, the staff will not in general make an independent assessment to the depth required for agreement with the detailed findings.

The review process will fall roughly into two phases. First, the staff will determine the completeness and adequacy of the documentation as submitted by the utility. This should be examination of the documentation to see that the requested level of detail has been provided for all subjects listed in Table 2.1.

Second, the staff will conduct a review of the content of the submittal, concentrating on event trees, system interactions and dependencies, failure modes, and treatment of containment function failure and radioactive material releases. This review constitutes a high-level sampling of the IPE. It is expected that specific fault trees will be requested by the staff during the review, the specific fault trees to be determined on a case-by-case basis. Questions directed to and meetings with individual licensees in order to clarify details of and discuss the examination process are to be expected.

The staff will review the options considered by the utility for plant improvements, including whether there are less costly alternatives if a utility found that there were no cost-effective options and whether there are any attendant risks associated with the proposed modifications. Further, the staff will review the list of vulnerabilities and the functional or systemic sequences selected under the screening criteria to obtain reasonable assurance that the licensee has made valid use of the insights concerning the plant.

An IPE Evaluation Report will be prepared documenting, in the same format as given in Table 2.1 for the utility's submittal, the staff review process and conclusions relative to the objectives in Section 1.1.

From time to time, the staff may find it necessary to perform a more detailed review or audit. The staff will determine the level of depth to which the detailed review should proceed, including independent assessment of parts of the IPE. It is likely that at least part of the documentation retained by the utility would have to be reviewed by the staff to accommodate the in-depth review.

If, as a result of its review, the staff determines that additional modifications appear to be warranted for a particular plant, it will follow the
procedure required by the backfit rule, 10 CFR 50.109.

At the end of the IPE process, the staff plans to prepare a document summarizing
its findings, insights, and conclusions relative to the goals of the Severe
Accident Policy Statement.

NRC FORM 335
(2-89)
NRCM 1102,
3201, 3202

U.S. NUCLEAR REGULATORY COMMISSION

BIBLIOGRAPHIC DATA SHEET

(See instructions on the reverse)

1. REPORT NUMBER
(Assigned by NRC. Add Vol., Supp., Rev.,
and Addendum Numbers, if any.)

NUREG-1335

2. TITLE AND SUBTITLE

Individual Plant Examination: Submittal Guidance

Final Report

3. DATE REPORT PUBLISHED

MONTH	YEAR
August	1989

4. FIN OR GRANT NUMBER

5. AUTHOR(S)

6. TYPE OF REPORT

Final Report

7. PERIOD COVERED *(Inclusive Dates)*

8. PERFORMING ORGANIZATION — NAME AND ADDRESS *(If NRC, provide Division, Office or Region, U.S. Nuclear Regulatory Commission, and mailing address; if contractor, provide name and mailing address.)*

Office of Nuclear Regulatory Research
Office of Nuclear Reactor Regulation
U.S. Nuclear Regulatory Commission
Washington, DC   20555

9. SPONSORING ORGANIZATION — NAME AND ADDRESS *(If NRC, type "Same as above"; if contractor, provide NRC Division, Office or Region, U.S. Nuclear Regulatory Commission, and mailing address.)*

Same as above

10. SUPPLEMENTARY NOTES

11. ABSTRACT *(200 words or less)*

Based on the Policy Statement on Severe Reactor Accidents Regarding Future Designs and Existing Plants, the performance of a plant examination is requested from the licensee of each nuclear power plant. The plant examination looks for severe accident vulnerabilities and cost-effective safety improvements that would reduce or eliminate any discovered vulnerability. This document delineates the guidance for reporting the results of a plant examination.

12. KEY WORDS/DESCRIPTORS *(List words or phrases that will assist researchers in locating the report.)*

IPE, Severe Accident Individual Plant Examination Vulnerabilities

13. AVAILABILITY STATEMENT

Unlimited

14. SECURITY CLASSIFICATION

(This Page)

Unclassified

(This Report)

Unclassified

15. NUMBER OF PAGES

16. PRICE

www.ingramcontent.com/pod-product-compliance
Lightning Source LLC
Chambersburg PA
CBHW081505170526
45166CB00008B/2559